ASTRONOMICAL TIDBITS

A LAYPERSON'S GUIDE TO ASTRONOMY

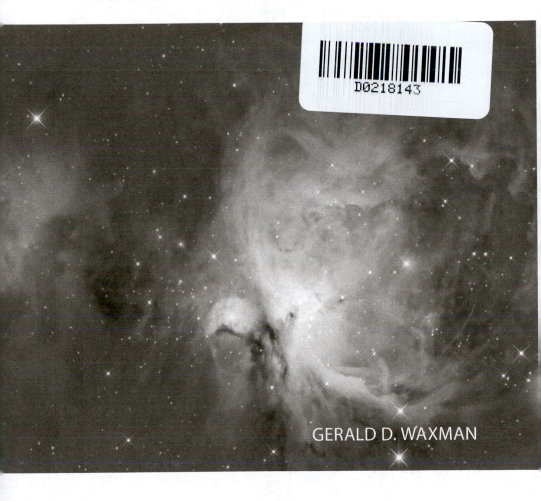

GERALD D. WAXMAN

Kendall Hunt
publishing company

Cover image © NASA

www.kendallhunt.com
Send all inquiries to:
4050 Westmark Drive
Dubuque, IA 52004-1840

Copyright © 2017 by Kendall Hunt Publishing Company

ISBN 978-1-5249-1354-0

Printed in the United States of America

Acknowledgments

To Greg Hayes, whose fine work and effort made this book possible. His work on the manuscript never ceased to amaze me.

Also, thank you to the other great friends who helped with this book: Ida Egli, Jane Schuler-Repp, Ken Holmes, Nan Van Gelder, Ron Smith, and Deanna Stropes.

Contents

Part 4—The Milky Way, Galaxies & Cosmology 123

List of Illustrations for Astronomical Tidbits

Preface

It is said that the history of astronomy and the history of science are one and the same. This is true! It is true because astronomy is the oldest of the sciences; it goes back thousands of years—to a time before there was writing. Standing stones (such as Stonehenge) and cave drawings clearly show our historic preoccupation with the heavens. Why this interest? I think the answer is twofold. First and foremost, astronomy is eminently practical. By looking at the sky, our ancestors were able to tell time, not only time of day, but also time of year. They could, therefore, predict when it was time for the sowing of seeds or the harvesting of crops. A second practical reason for looking at the sky was navigation. When you are traveling over unfamiliar terrain, you look for "landmarks" to guide your way. But there are no landmarks at sea and there are no "seamarks" either. Can you imagine the captain saying, "Okay sailor, remember that wave!" But there are "skymarks"—the stars and constellations—and observations of these will see you safely back to port (or, for that matter, sherry).

Another "practical" reason for watching the heavens is the belief that the seven celestial wanderers, for which we name the days of the week (Sun, Moon, Mercury, Venus, Mars, Jupiter and Saturn) are, either by conscious design or natural law, the driving force behind all Earthly happenings. In this pseudoscientific paradigm, known as "astrology," knowledge of the seven celestial bodies' positions relative to the background constellations should allow the practitioner to predict the future. Now that's practical! But is it accurate? That's where astronomers and astrologists must part company: whose observations are *you* willing to trust your life to? Let alone set your watch by?

But beyond the practical there is, I believe, a second, more fundamental reason why we look at the sky with wonder and longing—for the same reason that we stand, hour after hour, gazing at the distant swell of the open ocean. There is something like an ancient wisdom, encoded and tucked away in our DNA that knows its point of origin as surely as a salmonid knows its creek. Intellectually, we may not want to return there, but the genes know, and long for their origins—their home in the salty depths. But if the seas are our immediate source, the penultimate source is certainly the heavens.

You see, modern astronomy has demonstrated that the early universe was composed of only hydrogen and helium: all the heavier elements, more than 100 of them, were missing! Where did they come from? A century of careful sky watching has demonstrated that the rest of the elements were generated in the atmospheres of dying giant and supergiant stars. And further, after synthesis, these stars ejected these elements into the cosmos where they ultimately formed into many things, not the least of which was a planet called Earth. But that's not all. On the Earth there are living creatures—humans, for example—and these creatures are made out of potatoes, tomatoes, and spinach—all things made of "Earth!" The spectacular truth is—and this is something that your DNA has known all along—the very atoms of your body—the iron, calcium, phosphorus, carbon, nitrogen, oxygen, and on and on—were initially forged in long-dead stars. This is why, when you stand outside under a moonless, country sky, you feel some ineffable tugging at your innards.

Is it any wonder that the ancients were interested in astronomy? But a deep interest in the heavens does not insure a rational conclusion about the nature of the heavens—especially when one is convinced that the driving force behind celestial motions was the capricious desires of disembodied gods and goddesses. It was this reliance on superstition and religion that prevented our ancestors of 3,000 to 5,000 years ago from piercing the darkness of the long prehistoric night. After all, if the fall of every leaf or sparrow was the frivolous act of a capricious deity, what hope was there in being able to predict astronomical events?

But all this changed 2,500 years ago in ancient Greece. There, in the murky Aegean backwaters, arose a concept that was to change the world, a concept so powerful that nothing could stand against its might. Simply stated, the idea was that the natural world was *not* ruled by capricious forces. Instead, there were rules of nature—laws that demanded obedience. Yes, there might be gods and goddesses out there, but even *they* had to obey the rules. Moreover, and this is an important point, these laws of nature could, by suitable study, be discerned—we could figure them out. This concept is at the foundation of modern science.

Perhaps the first person who could be properly called a "scientist" was Pythagoras (6th century before the Christian era). Pythagoras and his Brotherhood of Pythagoreans believed that the essence of reality was in mathematical forms or "figures." This bears a striking resemblance to modern science, which believes that the best expression for scientific models comes through the language of mathematics. But along with their reliance on numbers, the Pythagoreans were thoroughgoing mystics, combining a religious fervor with their attempts to understand the world through numbers.

But Greek scientists faced one big, traditional problem. What did the term "suitable study" mean? In modern science, suitable study involves observation or experimentation. But to many ancient Greek scientists/philosophers, most notably Plato and Aristotle, suitable study meant resorting to authority. Because they felt the Earth was immobile and the planets had circular orbits, nobody seriously challenged their assertions. In fact, it was considered heresy to do so. So their words were passed on down the corridors of time and their errors were propagated. The peak of the disorder came in 150 A.D. when Claudius Ptolemaeus developed a system for predicting planetary motion. In order to make the orbits circular he needed to employ a series of off-center wheels. The net effect was a very complex theory which only worked marginally well. But the worst part of the theory was not that it was complex, nor that it only worked marginally well. It was dead wrong! And nobody dared try to correct it. These days, as soon as a theory is advanced, science colleagues try to prove it wrong. Back then, when resorting to authority was law, there were no checks and balances. The climate of the times became more and more anti-intellectual. In fact, by 450 A.D. a mob came to burn the library at Alexandria and kill the director, and there was no one to stop them.

And so the darkness descended. The brief flame that illuminated the Ionian world for a few precious centuries was extinguished and humanity fell into a fitful sleep called the Middle Ages that was to last for over 1,300 years. During that dark interlude, science, if it was practiced at all, was presided over by the ghosts of Plato, Aristotle and Ptolemy. The task of the scientist of the Middle Ages was not free inquiry. Inquisitiveness was discouraged. Rather, the task of the scientist was to read, interpret, and understand the teachings of Plato and Aristotle, and be able to use the Ptolemaic model to predict planetary motions so that horoscopes might be constructed.

We could continue on across centuries to detail step by step the observations and discoveries of Copernicus, Brahe, Kepler and Sir Isaac Newton who helped overthrow the antiquated models and bring us to the brink of our modern age of quantum physics, black holes and neutrinos, but to do so would require entire textbooks and time for us to read them. Instead, consider these "Astronomical Tidbits" I have assembled here. The next time you find yourself drawn outside to gaze at the night sky with wonder and longing, take this book along (with a flashlight!) to help you leapfrog through spacetime. The curious reader hungry for knowledge about the substance of Earth, Moon, Sun, the galaxies, even the "Search for Extraterrestrial Intelligence," will find in these pages an almost random dipping into the history of astronomy and the cosmos, to be experienced as one might consume bon-bons, or—for those looking to return to earthier roots—tomatoes, potatoes and spinach. What these "Tidbits" have to offer you is not so much a perfectly balanced meal as a scattering of all my favorite morsels, an offering I encourage you to consume as your appetite dictates.

PART ONE

Constellations & Sky Mythology

Chapter 1

The Big Dipper

Perhaps the most familiar grouping of stars is the one that we call the *Big Dipper*. Surprisingly, the Dipper itself is not a constellation. Rather, it is an *asterism*, a group or pattern of stars which is part of a larger constellation, or composed of stars from several constellations. In this case, the constellation of which the Dipper is part is *Ursa Major*, the Greater Bear (see figure). (An example of an asterism made of stars from several constellations is the *Summer Triangle*.)

The Big Dipper is quite handy for finding other significant stars, most notably the stars *Dubhe* and *Merak* (α and β Ursae Majoris), known as the pointer stars, which point the way to the North Star, *Polaris* (so-called because it lies almost directly above the Earth's North Pole), and, by association, to the *Little Dipper*, part of the constellation known as *Ursa Minor*, the Lesser Bear. Because it is very near the North Pole of the sky, as the Earth spins, all of the stars in the firmament move in circular paths around Polaris. These paths are called *Diurnal Circles*. You can watch this happening as the night progresses.

Interestingly, because the Earth is wobbling (precession), the axis describes a circle in the sky, and so Polaris is really the North Star only every 26,000 years. If we describe a circle every 26,000 years, how much do we move per year? The answer is .0138 degrees (one sixtieth of a degree is one arc minute). The reason for the precession is the gyroscopic action of the Earth. The Moon and planets are all acting on the tidal bulge trying to straighten the Earth, but its rotation causes it to remain tipped and the result is the motion of the pole.

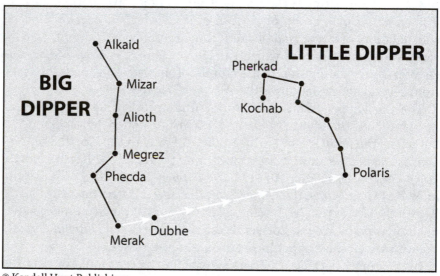

Chapter 2

Arc to Arcturus and Speed On to Spica

In the previous tidbit we discussed how Dubhe and Merak, the pointer stars in the Big Dipper, can be used to locate the North Star, Polaris. Now we will see how the Dipper can be used to locate another important star and constellation.

For decades, astronomers, professional and amateur alike, have used this short memory aid to find a significant neighboring star: "ARC to ARCturus." If you extend and follow the arc of the Dipper's handle, you will come to the bright star Arcturus.

Arcturus is a red giant star about 11 parsecs distant (1 parsec = 3.26 light years, so Arcturus is just over 36 light years from the Solar System). It is the brightest star in the constellation Boötes (pronounced "Buh-o-tez"), the bear driver, so-named because this group of stars follows Ursa Major, the Greater Bear, across the sky. In truth, Boötes looks far more like an ice cream cone (chocolate, of course) than it does like a bear driver.

Because Arcturus is the brightest star in its constellation, it is given the Greek letter alpha (α), followed by the genitive form of the constellation name. Hence Arcturus is also called Alpha Boötis. Arcturus and Boötes can be seen all summer high in the sky. Just "follow the arc . . ."

But it doesn't end here. The complete mnemonic phrase really is "Arc to Arcturus and Speed on to Spica." Spica is the brightest star in the constellation of Virgo. The star represents a "spike" of wheat, held in her left hand. The star is blue-white, which means that it is very hot (about 46,000 degrees Fahrenheit). It's distance is about 250 light years.

Virgo represents Persephone, goddess of the underworld and daughter of Zeus, and goddess of the harvest, Demeter. Persephone was so beautiful that Hades, god of the underworld, wanted her for himself and kidnapped her, taking her to his underworld. Zeus sent Hermes down to Hades to negotiate Persephone's release. Hades finally agreed but mandated that she return to the underworld for one-third of each year. The other two-thirds she could stay in the natural world with her mother. When Persephone was with Hades, her mother Demeter refused to let anything grow. Hence, the winter.

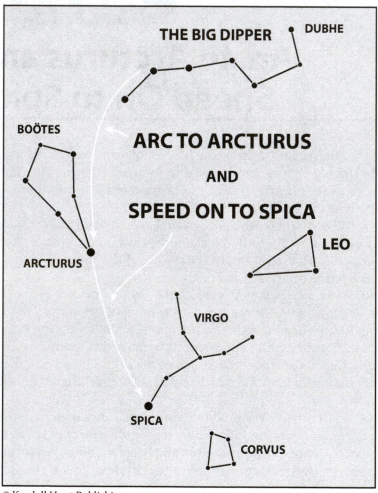

Chapter 3

Mysterious Sirius

Sirius has also been at the center of two mysteries. The first has to do with its color. We know Sirius as a white star of temperature 20,000 degrees Fahrenheit. Yet the Greek astronomer Ptolemy listed the star "fiery red" and others called it "redder than Mars." We know that red stars have temperatures of 6,000 to 8,000 degrees Fahrenheit. This implies that Sirius has undergone an extreme temperature change in the last two millennia. Astronomically, this is a very rapid change, not easily explained but by no means impossible. Of course, it is also possible that the ancient color descriptions were wrong.

A mystery concerning Sirius that has reached the popular press has to do with a remote African tribe called "the Dogon." From 1931 to 1952, French anthropologists worked with this tribe and, among many things, learned of their knowledge of astronomy. They found that the Dogon were deeply concerned with the star Sirius. They referred to a small, superdense companion to Sirius, which they said orbited the star every fifty years. Sirius does have a white dwarf companion, called Sirius B, that was first discovered in 1862, which orbits Sirius every 50 years. Some popular writers have suggested that the Dogon's knowledge of Sirius B is evidence that the tribe was visited by extraterrestrials from the Sirius system. Astronomers do not find this theory compelling, however. They counter that the stars in the Sirius system are not old enough to have spawned intelligent life. Further, Sirius B, in becoming a white dwarf, would have gone through a red giant phase, cooking the star system and any life that might have existed there at that time.

Another explanation for the supposed mystery is that the Dogon knew of Sirius through cultural contamination. This could have occurred during visits that were made by western missionaries in the 1920s or when Dogon tribesmen started attending French schools in the area in 1907.

Given competing hypotheses, the principle of Occam's Razor advises us that the theory that calls for the least assumptions is most likely to be correct. The competing theories here are that either the Dogon were visited by aliens from Sirius or that they learned of the astronomical properties of Sirius from westerners in the area. Which hypothesis do you think is most likely to be true?

Chapter 4

Algol: The Demon Star

Of all the stars in the sky, there is only one that varies in apparent brightness regularly. The star Algol, when watched closely, changes in brightness by a factor of 2.5 within a period of two days and twenty hours. What's going on here? The star is of a type we call an eclipsing binary. An eclipsing binary star is actually two stars orbiting each other, so far away that they look like a single star. In this case, the orbit plane of the stars is in the line of sight of the observer, so that one star periodically eclipses the other. When this happens, the brighter star is eclipsed by the fainter star and the system dims.

The star Algol is in the constellation of Perseus, hero of the story of Andromeda. Andromeda was the daughter of Cassiopeia and Cepheus, queen and king of ancient Ethiopia. Cassiopeia was a very vain woman. One day she compared her beauty to that of the sea nymphs, daughters of Poseidon. Poseidon, the god of the sea. Now, the sea nymphs overheard her boasting and complained to Poseidon, their father. To punish Cassiopeia, Poseidon created a sea monster, Cetus, who was given the task of destroying the coastal villages of Ethiopia. This he did with great relish. The residents of these villages complained to Cepheus about the situation. Cepheus visited his oracle to see what he should do. The response came quickly. Cepheus, to propitiate the monster, had to sacrifice his only daughter, Andromeda, to the monster by chaining her to rocks at the edge of the sea.

Meanwhile, in another part of the world, Perseus was just finishing a mission. It seems as though the people there were having a problem with a monster of another sort. Medusa was once a beautiful woman. But she, too, had a problem with vanity. Particularly, she had beautiful hair, which she compared to the hair of the goddess, Athena. Athena, in order to punish Medusa for her vanity, changed her hair into a nest of writhing snakes. This made her so ugly that whenever anyone looked at her, they were immediately changed to stone. The people in the village were getting tired of all the stone statues lying around, so they put out a contract on the head of Medusa.

Perseus was a hero. As such, he was in the rescue business—in this case, the business of rescuing the people from being turned to stone by Medusa. He went to the cave where Medusa slept. Brandishing the shield of Hera and the sword of her husband Zeus, he silently flew into the cave on winged shoes borrowed from Mercury. To avoid looking directly at Medusa and being turned to stone, Perseus used the shield as a rear view mirror, and cut off her head with one sweep of Zeus's sword. Realizing that the head still had its powers to turn people to stone, he put it in a sack and slung it over his belt. He then took off in further search of adventure. Unfortunately, drops of blood from the head of Medusa fell into the sea. Poseidon, who had loved Medusa when she was beautiful, saw what had become of her and became sad. He scooped up the drops of blood, mixed them with froth from the oceans of the world and sands from the beaches of Earth, and from this mixture he created a monument to her beauty, Pegasus, the winged horse. Pegasus sprang fully formed from the sea and greeted Perseus. They immediately became fast friends and went off together. They were flying over the coast of Ethiopia when they looked down and saw Andromeda chained to the rock, the monster Cetus bearing down on her. Perseus quickly landed on the rock in front of her, drew his sword and threw himself into battle with the monster. The monster was a creation of Poseidon and not easily defeated. Perseus was losing the battle. Suddenly he remembered the head on his belt, and withdrew it from the bag. The monster Cetus took one look at it, turned to stone and sank to the bottom of the sea. Perseus had rescued Andromeda.

All these characters—Perseus, Andromeda, Cassiopeia, Cepheus, Cetus and Pegasus—are constellations. The name Algol means the demon, and represents the head of Medusa on the belt of Perseus as he flies across the sky. In this way, the constellations are all connected to the ancient world and give us a way of seeing ancient culture.

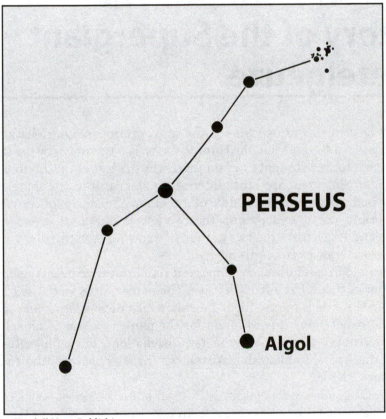

Chapter 5

Orion's Armpit: The Story of the Supergiant Betelgeuze

From December through March, the most prominent constellation in the southern sky is Orion, the Hunter. Orion is best known by the three stars that form a straight line and represent the belt of this mythological giant. But there are other interesting stars in this grouping. The three belt stars are in the middle of a rectangle made up (of course) of four bright stars (see diagram). The two lowest stars in that rectangle, Saiph and Rigel, mark the legs of the Hunter, while the two upper stars represent Orion's giant arms or shoulders.

If you look carefully at the upper left hand star in the rectangle, you will notice that it has a reddish tinge. The name of this star, *Betelgeuze*, is from the Arabic phrase Ibt al Tauzah, which means *the armpit of the central one (Orion).* It is common for the name of a star to reflect the function or position of the star in its constellation. For example, Rigel is from the Arabic Rijl Jauzah al Yusra, *the left leg of Jauzah* (the Arabic title for Orion).

The sanguine hue of Betelgeuze (this star is also known as Alpha Orionis) tells us that it is a very cool star. In fact, from knowledge of the precise color of a star, astronomers can calculate the precise surface (outer) temperature of that star. Betelgeuze is 3,000 Kelvin (about 6,000 degrees Fahrenheit) or half the temperature of the Sun, which is about 6,000 Kelvin. In the jargon of astronomers, such cool stars are called "M stars." Stars that are the temperature of the Sun are called "G stars."

The full range of temperature classes for stars are: O, B, A, F, G, K, and M, with the hottest being O and the coolest being M.

"Temperature Class" is alternatively called "Spectral Type" because the result of stars having different temperatures is that they have different appearing spectra. In fact, a star's spectrum is used by astronomers to infer the temperature of the star.

The apparently random letters that make up the temperature or spectral types are a result of the system having evolved over time. When mistakes were discovered in the original system (which was a far more orderly system), whole classes were switched or dropped entirely rather than switching the stars from group to group.

Nonetheless, depending on brightness and temperature, astronomers can calculate how large a star is. Betelgeuze turns out to be over 1000 times the diameter of the Sun! While this may seem large, astronomers have measured some stars to be as large as 1,700 solar diameters. Now that's big!

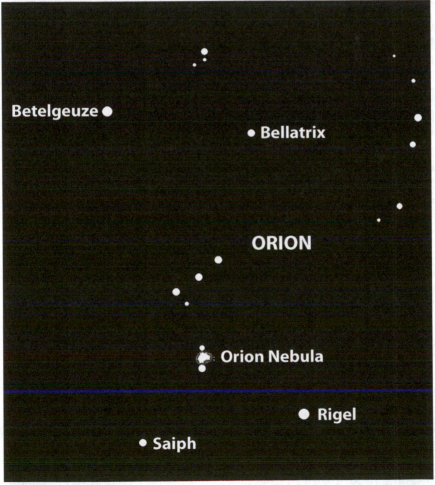

Chapter 6

Orion: The Hunter in Winter

One is apt to notice first the three stars in a row, which mark the belt. From right to left, *Mintaka, Alnilam* and *Alnitak* are their names and they mean, respectively, the belt, the string of pearls, and the girdle. Only later will you be told that the belt belongs to Orion, mighty hunter of the winter sky, and shown the other stars which, together with the belt, complete this brightest of all winter constellations. First there's Betelgeuze (α Orionis) marking Orion's right shoulder (left shoulder as you look at it), a brilliant red supergiant, whose intrinsic brightness is fully appreciated when you learn that the star is some 200 parsecs distant. That's over 600 light years (1 light year is approximately 6 trillion miles).

Across the belt from Betelgeuze, and equidistant from it, is the star Rigel (β Orionis), from *Rijl Jauzah al Yusra*, the left leg of Jauzah, the name by which the ancient Arabians knew the constellation. Rigel's blue color tells you that it is a hot star (T = 12,500 K) while its parallax (0.003 arc seconds) provides its distance. It is 333 parsecs away (or 1,100 light years), more distant than even Betelgeuze.

At the left shoulder we find Bellatrix, (γ Orionis), the female warrior or the Amazon Star. This giant star is at a distance of 40 parsecs (about 130 light years).

Finally, completing the rectangular outline of the figure is his right leg, Saiph, from *Saif al Jabbar*, the sword of the giant.

In Greek and Roman mythology, Orion was a giant, courageous man whose feats were rivaled only by Hercules. He was a favorite of the Gods and, being the son of Neptune, he could walk on water.

Despite all of this, Orion is a tragic figure, for he had the misfortune of falling in love with Merope, daughter of the island king Oenopion. Orion wooed Merope tenderly and patiently for many months and,

despite the fact that the maiden was willing, her father was uncon-
vinced and withheld his consent. Finally, because of the pressure that
Orion and Merope were putting on him, Oenopion said to Orion, "If
you will clear the island of all the wild beasts that infest it, so men can
walk from one end to the other without danger, then I promise you that
you shall wed my daughter."

Well, Orion loved hunting, and what's more, he was good at it. Before
very long he returned to Oenopion and told him that not a single wild
beast remained on the island. But Oenopion was still not satisfied. "Wait
a little longer," he told Orion.

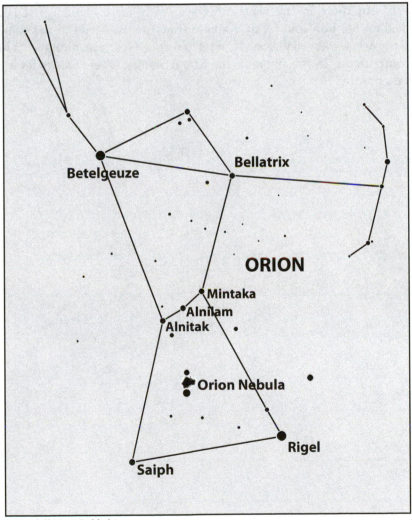

It was at this point that Orion realized the truth of the matter and that the king's promises meant nothing. Orion, in a drunken stupor, raped Merope whereupon Oenopion caused Orion to be blinded. Orion consulted an oracle who told him that he could be cured if he traveled east to the place that saw the first light of the new day and let the rays of the Sun strike his eyes. This he did and was cured. Whereupon he moved to Crete, where Diana, the goddess of the Moon, fell in love with him. So smitten was she with him that she failed to light the evening sky with moonlight. Orion died when Apollo, Diana's brother, became tired of his sister's affection for the mighty Hunter. He waited until Orion was swimming far out at sea and he challenged his sister to hit "that speck" out there in the waves. Always ready for a challenge, Diana picked up her bow and let fly a single shot. She was true to her mark, and the arrow pierced Orion's heart. Diana placed Orion, along with his accoutrements, in the sky near the Seven Sisters, where she visits him once a month.

Chapter 7

The Summer Triangle

When you use a star chart, you first need to identify one constellation or asterism that can serve as a starting point. In the winter sky, that constellation is usually Orion. In the summer sky, it is The Summer Triangle. The Summer Triangle is an asterism, made up of three stars from three different constellations that shine brightly high in the summer sky. The stars making up the triangle are Deneb, the tail of Cygnus, Altair, the head of Aquila, and Vega, in Lyra.

Cygnus represents a swan flying down the Milky Way with outstretched wings. The Milky Way in this part of the sky appears to be devoid of stars down its middle. This is actually an illusion. The place which appears to be devoid of stars is called *The Great Rift*. It is a cloud of dark, cosmic dust between us and the more distant stars. Significant astronomical objects within Cygnus are The North American Nebula, the bright double star, Albireo, and Cygnus X-l, the first black hole discovered.

The star Altair is part of the constellation of Aquila, the eagle, which can be seen flying parallel to Cygnus. This constellation is most famous as a home constellation of the star Altair 5, home star of the planet featured in the movie *Forbidden Planet.*

The constellation of Lyra represents the harp invented by Hermes and given to his half-brother, Apollo, who in turn gave it to his son Orpheus, the musician of the Argonauts (the sailors who sailed with Jason in search of the Golden Fleece). Vega, the brightest star in the northern hemisphere, is a star featured in the movie *Contact.*

Within The Summer Triangle are the minor constellations Vulpecula, the fox, and Sagitta, the arrow. Just outside of The Summer Triangle is Delphinus, the dolphin.

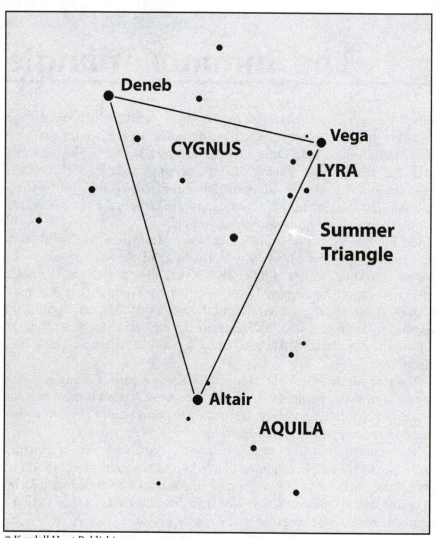

Chapter 8

Taurus and the Pleiades

One of the oldest constellations in the northern sky is Taurus. This constellation of the bull can be traced back almost 5,000 years. The Bull God, Apis is associated with Taurus. This Apis was no shadowy figure, but a living animal—a bull in whose form the divine Osiris was thought to be incarnated. The bull actually lived in Memphis, Egypt, where it was kept in luxurious surroundings and people from all over came to bring offerings to it. When chosen as a calf from the herd, the bull had to have a black tail with a white square on its forehead, a second white mark in the shape of an eagle on its back, and the circular lump on its back had to be similar in form to the sacred beetle or scarab.

After the true Apis had been recognized by all these signs, he was escorted down the Nile to Memphis in a splendid boat. Thereafter, the bull had an easy life. He was considered to be an oracle, offered food and every year on a date as officially fixed as his birthday, he was celebrated with special gifts and sacrifices. When the bull died, the entire country mourned his death and he was buried in a special mausoleum. Then the tradition began anew in the selection of a new Apis.

The constellation of Taurus looks like the front half of a Texas longhorn. The bull is emerging from the flowing river Eridanus, a faint southern hemisphere constellation. The tip of the southern horn of the bull is marked by a star that has the distinction of the longest name of any star in the sky. It is Shurnarkabtishashutu, which literally means in Babylonian, the end star of the southern horn of the bull. It is famous, for it lies one degree to the left of M1, the first object in Messier's catalogue for non-stellar objects. M1 is known as the Crab Nebula and is a remnant of the supernova of July 4th, 1054 A.D. The right eye (left to the observer) of the bull is Aldebaran, which means "the follower," because as the Earth turns it follows the Pleiades across the sky.

In the shoulder of the bull are the Pleiades, or Seven Sisters. They are a small cluster of stars resembling the Little Dipper and are often mistaken for such. The stars are actually about 700 in number but only six are bright enough to be seen with the naked eye. The seventh (the lost Pleiade) can be seen if you have exceptional vision. The Pleiades mythologically represent the daughters of Atlas, who were being pursued by Orion when Zeus saved them by putting them in the sky.

Chapter 9

The Heart of the Scorpion

One of the most beautiful constellations in the sky is Scorpius, the scorpion. Despite its beauty, the scorpion has always been identified with darkness and death. Mythologically, the scorpion was the animal that killed Orion by stinging him on the heel. But to be fair, it should be said that the stinging was called for by the goddess Juno, who became angered when she heard Orion boast that there was no animal on Earth that he feared. Even today, the scorpion is never found in the sky at the same time as Orion. One rises when the other sets.

The brilliance of the stars in this constellation is matched only by the stars in Orion. Scorpius has a curved tail and also pincers, if you include the stars that were robbed from the constellation Libra. The heart of Scorpion is the bright star Antares, which literally means *rival of planet Mars*, otherwise known as Ares by the Greeks. The bright red color of Antares signifies its cool temperature. The star is 3,000 Kelvin, or 6,000 degrees Fahrenheit. Antares is a supergiant star in the last stages of its life. It is almost 900 times the size of the Sun and 57,000 times as luminous. This makes it one of the most intrinsically brilliant stars in the sky. Its distance from Earth is about 600 light years.

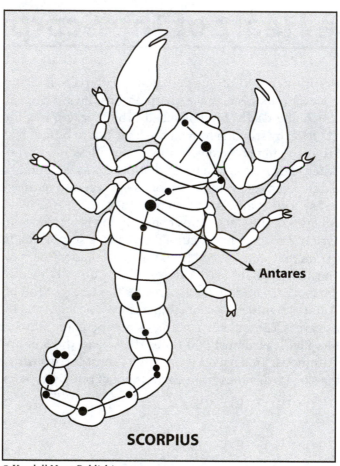

Antares

SCORPIUS

Chapter 10

The Pole Star and the Little Bear

The axis of rotation of the Earth undergoes a periodic wobble called "the precession of the equinoxes." This precession, analogous to the wobble of a top, is caused by the gravitational action of the Moon and Sun on the Earth's equatorial bulge. The effect is to cause the axis to point to different pole stars in different epochs. The period of the precession is about 26,000 years. Today the Pole Star is a star named Polaris.

The Pole Star, Polaris, can be found using the pointer stars, named *Dubhe* and *Merak*, which are the last two stars in the bowl of the Big Dipper. By "Pole Star" we mean the star that stands above the North Pole of the Earth. It just so happens that the Pole Star today lies at the end of the handle of the Little Dipper.

To the Native Americans, the grouping that we call the Little Dipper represented a little bear in the sky. The name of this constellation is Ursa Minor. The Big Dipper is part of *Ursa Major*, The Big Bear.

The constellation of Ursa Minor is very faint. The only stars bright enough to be seen easily are Polaris and the two stars at the end of the bowl. These two stars are called "The Guardians of the Pole." Named Kochab(β) and Pherkad(γ), these stars, at the time of Plato (400 B.C.) served as the Pole Stars. There is, as time goes by, a new Pole Star. Fourteen thousand years ago, the Pole Star was Vega. It will again be the Pole Star in 12,000 years. When Vega is the Pole Star, it will be correct to say that the Pole Star is an exceptionally bright star because it is the brightest star in the northern hemisphere. Polaris, our current Pole Star, by contrast, is the 50th brightest star in the sky.

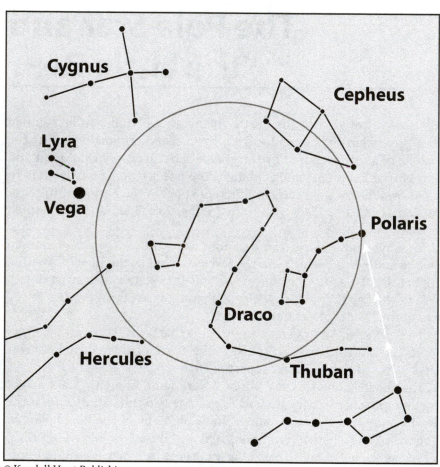

Chapter II

How Do We Know What We Know? A Close Shave With Occam's Razor

Have you ever met someone who clung to what you considered an irrational point of view despite your best efforts to set him or her straight? Just how does one "know" something? How do we know what we know? This is a very important scientific question, for it bears heavily on how a culture thinks.

Let us examine an example of a very hot topic: the existence of God. One has to agree that if one chooses to believe in the concept of a supreme creator, then his or her belief is backed up by no evidence one way or the other (and that's okay to do so as long as you know what you're doing). A belief is something you "know" to be "true" despite the fact that you have absolutely no physical evidence for it. Say you just have a "gut feeling" that something is true. It's okay to believe it, but you had better be prepared to be wrong. Because, depending upon what you're betting on, the odds of you being right (in the absence of physical evidence) may be long indeed. That is why such rigorous standards are applied to questions of scientific validity. When a scientist makes a statement of scientific fact, it (supposedly) has undergone the most rigorous testing possible. In fact, this is a tenet of basic scientific inquiry—that we must test to all reasonable limits any hypothesis that we want to accept. Not to do so is bad science.

Consider the God hypothesis. It is an unnecessary step to place God in any universe, for we can explain all we have to explain without recourse to God. And depositing a God is a violation of Occam's Razor.

What is Occam's Razor? It is a philosophical concept, which states that a simple hypothesis is more likely to be correct than a complex one. It is also known as the Principle of Parsimony. If you have two competing theories, logic tells you that the one that demands fewer assumptions is the one most likely to be true. To use a mathematical example, say you have a theory that you want to explain. Say one explanation has two parts, and each part has a probability of 50 percent of being right. The probability that the whole theory is right is 25 percent, or half of a half. In the second case, the theory has one part, which has the probability of 50 percent of being right. This explanation is the one preferred because it has only one part. And despite the fact that it has the probability of 50 percent of being right, it is twice as likely as the first case to be correct. Occam's Razor asserts that the more you multiply assumptions, the less chance you have of being right. The simplest explanation is the best.

PART TWO

Chapter 12

What is on the "Dark Side of the Moon"?

People often get confused with the terms Dark Side of the Moon and Far Side of the Moon. There is indeed a Far Side. There is no such thing as a permanent Dark Side.

The Moon is in synchronous rotation, which means that it rotates in the same period (time) as it revolves around Earth. Because the Moon rotates in the same period as it revolves, it keeps the same face toward us. The side of the Moon always facing the Earth is called the Near Side. We never see the backside of the Moon, known as the Far Side. Nonetheless, the Far Side is sometimes dark and sometimes lit. For example, when there is a new Moon, the Far Side is lit and the Near Side is dark. When there is a full Moon, the Far Side is dark and the Near Side is lit. So you see, all sides of the Moon take turns being dark and lit. At quarter Moon, half of the Near Side is lit and half of the Far Side is lit; the other halves are dark.

There is one time when a portion of the Dark Side of the Moon is lit; that is when there is Earthshine. Have you ever seen the dark part of the Moon glowing dimly? This occurs in the crescent phase just before or after new Moon. You can barely see the Dark Side glowing. This occurs because of sunlight reflecting off of the Earth and illuminating the Moon's dark hemisphere. If you were on the Moon you would see a bright, almost full, Earth in the sky illuminating the moonscape around you. This is analogous to being on the Earth and having a full or almost full Moon in the night sky. The moonlight lights up the surrounding landscape. I guess you would have to call this phenomenon "Moonshine," not to be confused with the more well known definition of "moonshine."

Chapter 13

Have You Seen "The Moon Illusion"?

You have seen the phenomenon of a rising full Moon appearing enormous in the sky. The interesting thing about this phenomenon is that the apparent size of the Moon is illusory. Yes, that's right. The Moon really is not bigger at that time. In fact, it is smaller than when it is high in the sky. The illusion is in your head, as can be proven by changing your perspective on the situation.

If you look at the Moon through a paper towel tube or bending down and looking through your legs, the Moon will revert to its proper size. The mystery of the Moon Illusion has plagued philosophers and scientists for two thousand years, and well it should. It turns out that the phenomenon is physiological, not physical. There is something that goes on between your brain and your eyes having to do with the juxtaposition of the horizon and the Moon. Beyond that, there is no explanation.

I said before that the Moon is actually smaller when it is on the horizon than when it is high in the sky. This is because it is four thousand miles farther away when it is on the horizon. Four thousand miles is the radius of the Earth. The diagram illustrates the difference in distance between viewing the Moon directly overhead (point A) and viewing the Moon as it sets on the horizon (point B). This difference is equal to Earth's radius (point C).

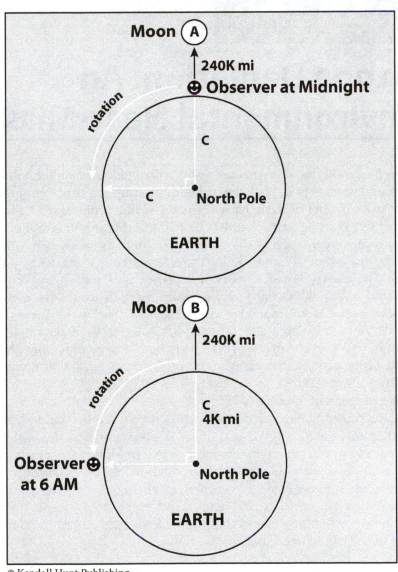

Chapter 14

Ozone Depletion: An Environmental Near-Miss

It has been said that everyone talks about the environment but nobody does anything about it. Well, that is not quite true. The site of our greatest victory in environmental science is the ozone problem. More than twenty five years ago, two chemists named M.J. Molina and F.S. Roland wrote a paper showing how the chemical Freon in its native form would destroy the ozone layer of the Earth. It was roundly attacked by the chemical industry, but after years of experimentation scientists proved that they were right. Ozone levels were dropping all around the globe.

Now, ozone is an important component of the Earth's protective shield, for it intercepts and blocks ultraviolet (UV) radiation from reaching the Earth's surface. This UV is high-energy radiation which could be responsible for many cases of skin cancer, not to mention death of phytoplankton in the Earth's ocean. These plankton are the source of much of the Earth's oxygen.

It was shown that the air above the South Pole was becoming highly deficient in ozone, putting Australians at great risk, and the problem was getting worse. Finally, ninety nations met in Montreal, Canada and signed an agreement to limit the production of fluorochlorocarbon compounds such as Freon. Now different chemicals take the place of the old ones and serve as refrigerants and propellants for spray cans. The important point to see is that humans saw the problem and took steps to solve it. Now on to global warming.

Chapter 15

Passage Into Darkness: Lunar Eclipses

Total lunar eclipses are interesting if not spectacular events. While solar eclipses are highly sought after, their poorer cousins are hardly noticed, except, perhaps, by astronomers. As with all eclipses, the business at hand is shadows, in this case the shadow of the Earth. Lunar eclipses always take place at full Moon. At this phase we have the Earth between the Sun and the Moon. If the alignment is perfect, the Moon can pass through the Earth's shadow, and when it does, the full Moon, momentarily deprived of its sunlight, goes dark, but not entirely dark. Some light creeps around the Earth, refracted through the atmosphere to illuminate the otherwise darkened Moon. So, in fact, the totally eclipsed Moon appears coppery red.

The frequency of lunar eclipses is related to the frequency of solar eclipses. An eclipse season is the time when the Sun is near enough a node (the place where the Moon's orbit around the Earth intersects the Sun's path through the sky) for there to be an eclipse (if there is a full Moon). The eclipse seasons are the same, twenty-four days for total eclipses and thirty-one days for the lunar eclipses of any type. As you can see, the lunar eclipses must occur during every eclipse season, whereas the total eclipses do not have to occur.

Lunar eclipses are divided up in a similar manner to solar eclipses. They are characterized as penumbral if the Moon moves through the penumbra of the Earth's shadow, and umbral if the Moon moves through the umbra. Umbral eclipses are divided into two types: total and partial. Penumbral eclipses darken so slightly that it is difficult to tell that the Moon is in eclipse. And so it is usually ignored.

Chapter 16

Phases of the Moon: What Are They?

We all know that the Moon operates in phases. But do we know what causes them? The phases of the Moon are quite simple, yet grossly misunderstood. They are caused by the fact that the Moon is always half illuminated by the Sun and has no light of its own. We, on the Earth, see different portions of the illuminated half. When we see none of the illuminated half we call that a new Moon. When we see the entire illuminated half we call that a full Moon.

It is a common practice to start the phases of the Moon at new Moon (see diagram). Each day the Moon moves twelve degrees around the Earth (this comes from the fact that the Moon moves once around the Earth, 360 degrees, in one month, about 30 days, and 360 divided by 30 equals twelve). At the end of a week, the Moon has moved almost 90 degrees around the Earth and is in the position marked A on the diagram. This is called first quarter Moon and we can see what looks like half a Moon in the sky. So if it looks like half a Moon, why don't we call it a half Moon? Because the Moon has gone a quarter of the way around the Earth in its orbit, starting from new.

The phase between new and first quarter is called crescent and it is obvious why this is so—because the Moon looks that way. During this time, the Moon looks like a sliver of a fingernail or top of a crescent wrench.

After first quarter Moon but before the full Moon we have a phase called gibbous (see diagram). This phase is least known. The word gibbous comes from the Italian word meaning hunchback, because the Moon looks hunched over. After gibbous comes full Moon. This is the best-known Moon. This Moon is so well known that it even has a monthly name. For example, the full Moon that occurs nearest the September

(Autumnal) Equinox is called the Harvest Moon, and the full Moon that occurs in the month of October is called the Hunter's Moon or Blood Moon. The full Moon is always opposite the Sun in the sky (as can be seen from the diagram) and therefore always rises at sunset and sets at sunrise. In fact, it is possible to tell time from the Moon. For example, if you see the full Moon rising you know it is about sunset, and if you see the full Moon setting you know it is about sunrise.

At this point we should mention the phenomenon which sometimes occurs at full Moon, and that is the Red Moon. This is not the same as the Blood Moon, although the Blood Moon may have been first named because of the Red Moon. The Red Moon occurs because the rising full Moon is low on the horizon and the atmospheric effect causes it to be reddish, much as the setting Sun and rising Sun are reddish. This effect occurs because of atmospheric extinction (when the atmosphere scatters certain wavelengths of light out of a beam). For example, the atmosphere scatters blue light preferentially, so a beam of sunlight looses its blue light before any other color (the blue light which is scattered is why the sky is blue). On the other hand, the red light is least scattered and will tend to stay in the beam longer than the blue. This is why the setting Sun is reddish and it is also why the rising Moon is reddish. It is just the thick atmosphere that lies between you and the Moon.

Now the phase is repeated in reverse. After full Moon you have a waning gibbous (the phases before the full Moon are called waxing). The waning gibbous rises rather late at night and sets early in the morning, and as such is not generally seen.

The end of the gibbous cycle terminates the third or last quarter Moon. This Moon looks like the first quarter Moon except that it is the other half of the Moon that is illuminated (at first quarter it is the western half of the Moon that is illuminated and at third quarter it is the eastern half). The way to tell the third quarter from the first quarter is that at first quarter the right side of the Moon is lit up while the left side is dark. At third quarter the left side is lit and the right is dark.

After the third quarter comes waning crescent which leads us again to new Moon, and the end of the cycle. A point of interest must be noted here. There is a widely held misconception that the Moon does not rotate. Actually, it does. It rotates one time per revolution and this synchronous rotation keeps one face of the Moon turned perpetually to the Earth (the reason for the misconception).

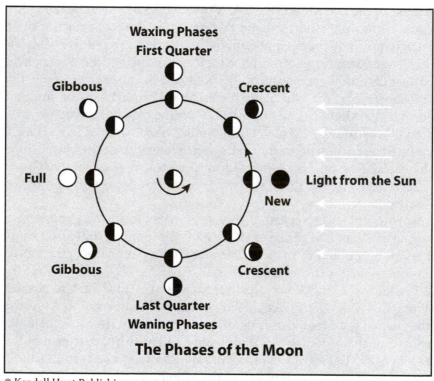

The Phases of the Moon

Chapter 17

Plate Tectonics: Will the Giants and Dodgers Again Be Cross-Town Rivals?

Have you ever noticed how a map of the world looks like a jigsaw puzzle? The east coast of South America tucks into the west coast of Africa and Europe tucks into the Gulf of Mexico. Well, the continents of the world are all in motion, sliding around on the surface of the Earth like blocks of ice floating in the water. When this theory was first presented it was called continental drift and was roundly ridiculed. We now know that the surface of the Earth is divided into crustal plates and the theory has a new name called "Plate Tectonics." It is a demonstrated fact that these plates are shifting at an average rate of one inch per year over the whole surface of the Earth. Where the plates come in contact with each other we get such phenomena as earthquakes, volcanoes, mountain-building, and ocean trenches. The crustal boundaries can form in several ways. You can have them converging toward one another, in which case you will get a mountain as one plate slides over another. You can get these plates diverging or moving away from one another, in which case you have an ocean basin. Or you can have these plates moving against each other as in the case of the San Andreas Fault, where one plate is moving north and one plate is moving south. The whole of the United States is on one plate called the North American Plate, except for the far west coast. San Francisco is on the North American Plate. Los Angeles is on the Pacific Plate. San Francisco is moving south while Los Angeles is moving north. The rate is one inch per year. In about 22 million years the two cities will be together.

Proof of plate tectonics came when, after World War II, geologists were mapping the magnetic field of the Earth by dragging magnetic probes across the ocean floor. They found, to their surprise, that the magnetic field changed direction periodically as they dragged the probe across the sea floor. There were magnetic stripes across the floor of the ocean, north-south, south-north. What was happening is that in the middle of the Atlantic Ocean was a spreading center, where new material was welling from the mantle of the Earth. This warm molten material came out of the mantle and spread outward in both directions, east and west, at the Mid-Atlantic Ridge. The warm material cooled and solidified, capturing the extant magnetic field which changed periodically and giving rise to the magnetic stripes that had been discovered. Spreading centers like in the mid-Atlantic are places where the crust is born. The crust dies where it dives under another plate, there to be subducted and remelted, becoming new mantle. In this way the Earth is constantly renewing itself. The surface is born at the mid-oceanic ridges and dies in the subduction zones. Along the way the crustal plates carry the continents on their odyssey.

Chapter 18

Sun at Latitude Zero: The Vernal Equinox

Every March 21ˢᵗ something special happens in the world. It is that day that we call the "Vernal Equinox," or "first day of spring." Now "the first day of spring" is a term that has no meaning, but the term "Vernal Equinox" has a very special meaning. The word "equinox" means "equal nights," for it is on this day that the Sun in its yearly trip around the Earth crosses above the equator. Every place on the Earth for one day has twelve hours of night and twelve hours of day. Now let me explain. You all probably know the Earth goes around the Sun once a year and not vice versa. So what is standard practice in astronomy is to look at it from the point of view of the Earth: in a geocentric perspective. From this point of view, the Sun revolves around us once a year. It really makes no difference which way you look at it.

Because the Earth is tipped on its axis, the Sun is above different places in different seasons. For example: the Sun is above latitude 23.5 degrees north in June (Mexico) and above minus 23.5 degrees (Namibia) in December. The Sun is above the equator in March and September. So while we know that actually the Earth is doing the moving, we may imagine the Earth to be at rest and the Sun to be in motion. Thus the Sun, once a year, revolves around the Earth, a mirror image of the Earth orbiting the Sun.

Now, the orbit of the Sun around the Earth is tilted by an angle exactly as much as the Earth is tilted, 23.5 degrees. So the Sun in its trip around the Earth goes from plus 23.5 degrees to minus 23.5 degrees, crossing the equator once on its way south and again on its way north. March 21ˢᵗ, the date of the Vernal Equinox, heralds the crossing of the equator by the Sun as it goes from south to north and 6 months later, on September 23ʳᵈ, the day of the "Autumnal Equinox," the Sun is crossing the equator again on its orbit southward.

Generally, there is unequal daylight on the Earth. When the Sun is north of the equator, the northern hemisphere has more hours of daylight than darkness. Correspondingly, when the Sun is south of the equator, the the northern hemisphere has more darkness than light. On the dates of the equinoxes, the day and night equalize. These are the only two dates of equality of day and night at every location on the Earth.

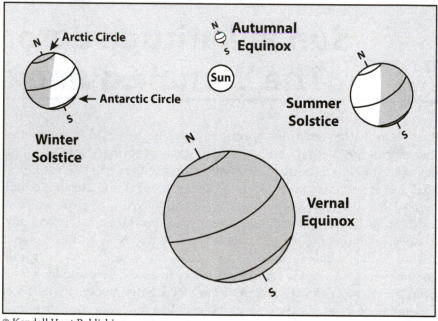

Chapter 19

Thank Heavens For the Greenhouse Effect: Without It We'd Be Cold Toast

Everybody should know about climate change. Because of all the carbon dioxide we are injecting into the atmosphere, the global average temperature of the Earth is predicted to rise. Indeed, it is rising already. This is an example of what we call the Greenhouse Effect, the trapping of radiation by certain gases in the atmosphere.

Just what is the Greenhouse Effect? It is like when we walk into a greenhouse and the temperature is higher than outdoors. This is true because the visible light entering the greenhouse through the glass is changed to infrared by the objects in the greenhouse. For example, a plant will absorb visible light and actually heat up. It will then re-emit the light in the infrared (heat) region of the spectrum. The glass, while perfectly transparent to visible light, is black so far as the infrared is concerned. In effect, the glass allows light in but not out.

In our atmosphere there are certain greenhouse gases that cause a warming of the atmosphere. The gas with which we are most familiar is carbon dioxide, emitted by the burning of fossil fuels. What is not so well known is that there are other greenhouse gases. Chief among them is water. Water vapor in our atmosphere also causes a greenhouse effect, which warms the Earth. Were it not for this effect, the average temperature of the Earth would be zero degrees Fahrenheit and the ocean would be frozen solid. No life could exist under these conditions. So the Greenhouse Effect is a good thing. Except when taken too far.

Chapter 20

The Catastrophic Origin of the Moon: A New Theory

Just how was the Moon formed? In general, there are three ways that satellites form. The most common way is accretion. In this method, the satellite undergoes gravitational condensation out of material left over after the formation of the planet it orbits. Satellites formed in this way generally orbit the equator of the primary planet. They are also generally made of the same substance.

Another way satellites form is through fission. Fission is a planet splitting apart, usually due to rapid rotation. Here, too, the satellite orbits the equator of the planet and is made of the same substance. The third way a satellite can be formed is through capture. In this method a satellite-to-be is orbiting the Sun in its own pathway, separate from the planet. It wanders in too close to the planet and is captured by its gravity. This kind of satellite can orbit the planet in any direction and usually has a different composition from the planet because it comes from a different place in the Solar System.

Of the three methods, the least likely to happen is capture—because of energy considerations. When one gravitational body falls into the gravitational influence of a second body, it always speeds up. So when the two bodies are as close as they are going to get, the fast-moving body has escape velocity from the second body, and therefore it passes through the gravitational influence. So in order for one body to capture another, it must rob the approaching body of its energy. An example of this is in order here.

In the U.S. Space Program, we sent six teams of men to the Moon. When they reached the Moon, their spaceships had to be slowed down in order to stay in orbit around the Moon rather than passing through its gravitational influence. So retro-rockets were fired when the spaceships were at their closest approach to the Moon. This slowed them down enough to stay in orbit around the Moon. If retro-rockets had not been fired, they would have come back to the Earth directly. This was the case with Apollo 13, the sole failed mission.

Ordinarily, capture would be our best method for explaining the origin of the Moon. This is because the Moon's orbit is more closely aligned with the plane of the Solar System (Ecliptic) than it is with the equator of the Earth. This argues strongly that the Moon suffered a random capture from somewhere else in the Solar System. The Moon's orbit is tipped by only 5 degrees from the plane of the Solar System, whereas the equator is tipped by 23.5 degrees from the plane of the Solar System.

Before we went to the Moon, the best theory of its origin was accretion. But there were problems with this theory. The Moon does not orbit the equator of the Earth; rather it orbits between 18 to 28 degrees of the equator. Also, Moon rocks brought back from six missions show that the Moon is, in general, like the Earth in composition. But it is different in one major point: the Moon is highly deficient in iron, compared to the Earth. This is difficult to explain in terms of an accretion hypothesis. The Moon's orbit is more closely aligned with the plane of the Solar System than is Earth's equator.

In recent years, a new theory has popped up to explain the origin of the Moon. This is the collision hypothesis. In this theory, the Moon was once a Mars-sized body orbiting the Sun elsewhere in the Solar System. It collided with Earth, smashing into the Earth's crust. After the collision, some of the invading body's material, along with Earth material, was blown out of the Earth. Some material was left embedded in the Earth, and a Moon-sized portion of both bodies ended up falling into an orbit around the Earth. This theory explains all the salient observations. It explains why the orbit of the Moon is tilted from Earth's equator. And it explains the similarities as well as the differences in composition of the Earth and Moon.

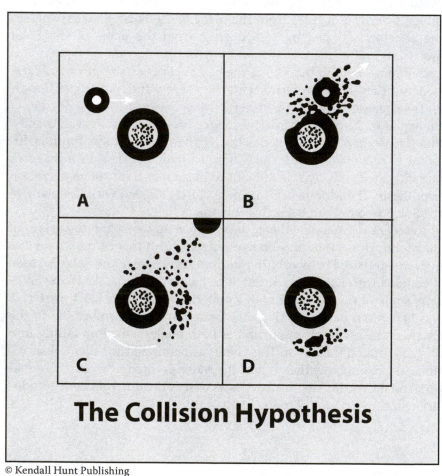

The Collision Hypothesis

Chapter 21

The Moon: When Will It Pop Up Next?

The Moon appears at seemingly random times. Just when you least expect it, there it is! This would make you believe that one cannot predict when the Moon is going to appear. In actuality, the rising of the Moon is quite predictable, as is its phase, such as a crescent or a quarter Moon.

When we say the Moon is rising, it is actually the Earth that is turning and bringing the Moon into view. So, say on day one the Moon rises at 6 a.m. What time will it rise on day two? The Moon moves in an eastward direction, 360 degrees once around the Earth every 30 days and therefore, 12 degrees every day. On day two, the Earth has to rotate an additional twelve degrees to catch up with the Moon. How long does it take to rotate 12 degrees? The Earth turns 360 degrees every 24 hours, so it turns at 15 degrees an hour. If one hour is 15 degrees, 12 degrees is approximately 50 minutes. So on average, each day an observer on Earth has to wait an additional 50 minutes for the Moon to rise. On our day two, the Moon will rise at approximately 6:50 a.m. and day 3, at 7:40 a.m. And so it progresses through the month until day 31, when we are back where we started on day 1 at 6 a.m.

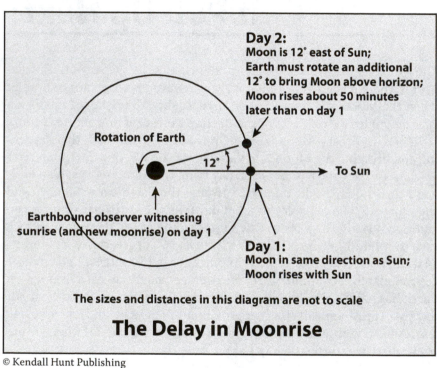

Day 2:
Moon is 12° east of Sun;
Earth must rotate an additional
12° to bring Moon above horizon;
Moon rises about 50 minutes
later than on day 1

Rotation of Earth

12°

To Sun

Earthbound observer witnessing
sunrise (and new moonrise) on day 1

Day 1:
Moon in same direction as Sun;
Moon rises with Sun

The sizes and distances in this diagram are not to scale

The Delay in Moonrise

Chapter 22

Meteors and Meteor Showers

What could be more exciting than a bright meteor on a moonless night? The answer: a meteor shower on a moonless night. Meteor showers can display many dozens of meteors per hour. They are rare but not so rare. One occurs every month. Meteor showers are an interesting story. To begin with, a meteor is a flash of light made by a falling meteoroid. The meteoroid enters the atmosphere at many miles per second. It's velocity causes intense friction with the atmosphere, causing it and the air to heat up. The air glows, the meteoroid glows and we see a streak of light in the sky.

On certain nights throughout the year the hourly rate of meteors, which on any one night is about 7 per hour, rises to up to 50 per hour. Thus, we have a meteor shower.

Meteor showers are identified by a radiant point which is simply a direction in space from which the meteors appear to be coming. You see, the meteoroids are all traveling along parallel paths in space and we know that, by perspective, parallel lines appear to diverge from a point in the distance.

This gives the effect of having the meteors diverging from a point in the distance, the radiant point. The different meteor showers are named after their radiant points. For example, on the night of August 12th every year we have a meteor shower whose radiant point is in the constellation Perseus. This is the Perseid shower and it sports about 50 meteors per hour. It is usually the best shower of the year. Likewise there is a Leonid shower whose radiant point is the constellation Leo. This meteor shower is on November 16th or 17th. Similarly there are the Geminids and the Orionids. In all, there is about one shower per month.

© Kendall Hunt Publishing

When should you go outside to see a meteor shower? The answer is, the later the better: after 3 a.m., and before dawn is a good rule of thumb. The reason is that between 3 and 6 a.m. you are on the side of the Earth that is rushing forward in space, the so-called leading edge. Earlier, around 9 p.m., the observer finds him or herself on the trailing edge, the back side of the Earth. Just the way your front windshield has more dead bugs than the rear windshield, so the leading edge of the Earth gathers more meteors.

An astronomy professor once gave a class a question about the difference between a meteor and a meteorite. One student was something of a poet and answered the question this way:

> *"A meteor is a flash of light made by a falling meteorite while rushing through the air in flight, usually seen at night. I hope to God this answer;s right."*

Within the limits of poetic license, it was and the student received full credit.

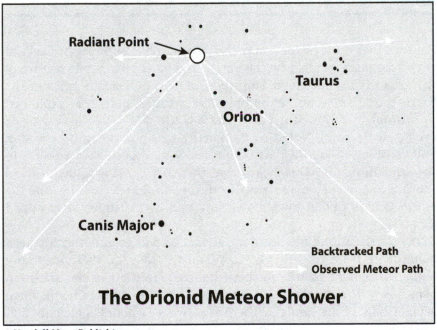

Radiant Point

Taurus

Orion

Canis Major

Backtracked Path
Observed Meteor Path

The Orionid Meteor Shower

Chapter 23

The Midnight Sun: Days Without End

I'm sure that most of you know that the number of hours of daylight vary from place to place on the Earth. The most remarkable feature of this effect is the midnight Sun: the fact that on certain parts of the Earth, the Sun does not set for months at a time. This can be explained by a simple diagram. The Earth's axis is tilted by 23.5 degrees to its orbit (see diagram). Note that the Earth's axis points toward the star, Polaris, always. So as the Earth orbits the Sun, the axis always points in the same direction. This means that the North Pole is tipped toward the Sun in one season and away from the Sun in another. In June the North Pole is tipped toward the Sun, and in December it is tipped away.

In June the Sun shines down on latitude 23.5 degrees in the northern hemisphere from directly above. When this occurs, there are more hours of daylight in the northern hemisphere than in the southern hemisphere. In December, this is reversed and the southern hemisphere gets the bulk of the sunlight. There are always 12 hours of daylight and 12 hours of darkness at the equator. The diagram demonstrates that the midpoint of the line of the twilight zone of dawn and dusk falls year round at the equator. At this midpoint, there are always equal hours of daylight and darkness.

During the summer in the northern hemisphere, the farther north you go, the more daylight you find. As you travel north, you reach a point where there are 24 hours of daylight. This is defined as the Arctic Circle. Anywhere between the Arctic Circle and the North Pole, the Sun does not set on June 21st. This is because the area within the Arctic Circle has moved completely into the daylight side of the twilight zone and the area within the Antarctic Circle has consequently also moved completely into the dark side of the twilight zone.

When you are at the North Pole, the Sun rises from a six-month sleep on March 21ˢᵗ. It spirals into the sky to reach an altitude of 23.5 degrees on June 21ˢᵗ. All of this time, the Sun is up and then it starts slowly spiraling down, to disappear entirely for another six months on September 22ⁿᵈ. This timing is just the opposite at the South Pole.

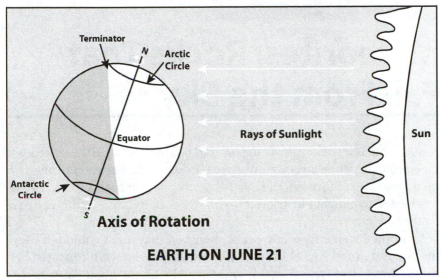

Chapter 24

Meteorites: Rocks That Fall From the Sky

A meteorite is a rock from space on Earth. This distinguishes it from a meteoroid, which is a rock from space, in space. In between times, the meteoroid passes through the Earth's atmosphere at high speeds, where it is heated to glowing by friction—the streak of light created is called a meteor.

Until the Renaissance, few people believed that rocks could fall from the sky. But in the Age of Enlightenment, those who were educated began to think otherwise. The tide was turned with the great Siena meteorite fall, where over 200 stones from the sky fell on the town of Siena, Italy. Then people had to believe the truth. Meteorites from the sky come in three varieties, one of which are the irons—meteorites made of iron and nickel. Although the irons are the most commonly found type of meteorite, the most common to fall are stones made of silicates. The reason we know this is that observed falls are mostly stones. Meteorites of a third type are stony-irons. All meteorites are traceable back to the Asteroid Belt, where they originated.

The most famous fall in our part of the world is the Canyon Diablo fall of about 50,000 B.C. At this time, a mostly iron meteoroid (small asteroid) of about 100-150 feet in diameter crashed into the Arizona desert. Most of the meteoroid was vaporized—only about 10 percent survived. Today, you can visit the site and find a large crater. It is called Meteor Crater, and it is found between Flagstaff and Winslow. It is the best preserved meteorite crater in the world.

If rocks fall from the sky, is there any danger of being hit? Well, there's only one case of a person being hit by a falling meteoroid/meteorite. In 1954, Ann Hodges of Sylacauga, Alabama, was taking an afternoon nap when a nine pound iron meteorite crashed through the roof of her house. It bounced off a table radio and hit her in the hip.

She received a nasty bruise. It's interesting to note that a lawsuit developed over this case. You may wonder who you sue if you're hit by a meteorite. Well, in this case Mrs. Hodges was renting the house. The lawsuit developed over claim to the meteorite. Her landlord, a Mrs. Lewis, sued for the rights to the meteorite. She claimed it was hers, because it was her house. Mrs. Hodges claimed that it was hers because it was her hip. The case was settled by the University of Alabama, who stepped in and offered both parties compensation for their loss, and kept the meteorite for itself. It now resides in the university museum.

Chapter 25

The Day the Sun Stood Still: The Summer Solstice

Every June 21st the Sun stands still. Actually, it doesn't really stand still—it's just what we say. June 21st is the date of the Summer Solstice—the stationary Sun. It is on this day that the Sun traveling northward in our sky apparently stands still, then reverses direction and travels southward. Every day the Sun apparently moves one degree around the Earth. Really though, the Earth moves one degree around the Sun. But on Earth we see the Sun as moving.

This day also marks the first day of summer, arbitrarily chosen. The first day of summer could have been any day. But this day was chosen because it has the most hours of daylight for a northern hemisphere observer. For a southern hemisphere observer, June 21st has the minimum in sunlight hours. This is because the Sun is shining down on the Earth from above the northern hemisphere. Another way of saying this is that on June 21st the sub-solar point on the Earth is in the northern hemisphere—at twenty three and a half degrees north latitude, to be precise.

All this should be explained. The Earth is tipped by twenty-three and a half degrees in its trip around the Sun. This means that the sub-solar point oscillates between minus twenty-three and a half degrees and plus twenty three and a half degrees. So on the 21st, we see the Sun as turning southward and we know that if you are going in one direction and want to reverse that, you have to stop first. Hence, the Solstice (which means "stationary Sun").

Because the Sun is shining down on the northern hemisphere, the North Pole is always illuminated as the Earth turns. This gives rise to the phenomenon known as the Midnight Sun in the northern hemisphere (see the essay entitled "The Midnight Sun: Days Without End"). In fact, the further northward an observer lives, the more hours of daylight he or she will have until arriving at the Arctic Circle. From the Arctic Circle to the North Pole, you have full daylight on June 21st.

All this is reversed for the southern hemisphere. There, the least hours of daylight are on June 21st, and the most are on December 22nd.

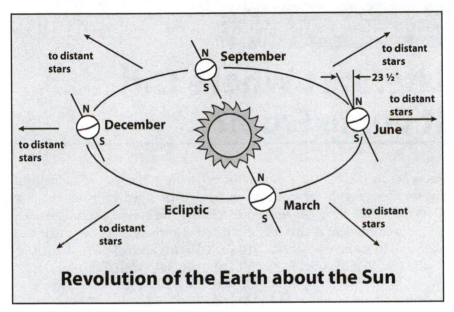

Revolution of the Earth about the Sun

Chapter 26

Life: Just Where Did It Come From?

There are two thoughts about how life originated. That is, two thoughts if you leave out the divine explanation, as scientists generally do. The first thought is that life originated on the Earth and spread out from there. The second thought is that life originated elsewhere, and was carried to the Earth by some mechanism. We shall discuss each of these in turn.

The idea that life originated on Earth is fraught with difficulties. How you make the first living entity, for example a DNA strand, is difficult to imagine if it occurred by trial and error. That's because a DNA strand is very complex, consisting of many thousands of atoms, all being placed in a proper sequence. It's difficult to see how life could have gotten to this point by random trial and error. Of course, time is our friend, and given enough time all things are likely, even unlikely things.

Another problem with this hypothesis is that it would seem difficult to evolve a functioning organ without knowing what it is before you start. For example, an eye or a heart—things that won't work unless they have all their components in place at the beginning. Science is still wrestling with that one.

On the other hand, we have the theory that life was transported to the Earth from afar. Astronomers have hypothesized that life was carried to the Earth by meteorites and comets, also supplying water to the Earth.

Other scenarios exist. Maybe spores, floating freely in the gulfs of space, filtered down onto the Earth along with our constant supply of micro-meteoric dust.

One thing is for sure. If life is ever discovered elsewhere in the universe, and it is based similar to Earth life, there will be a strong argument for a foreign origin of life. For only in that way will the odds of a dual creation be brought into respectability.

Chapter 27

Fire in the Sky: The Day Life on Earth Almost Ended

Everybody knows what a dinosaur is, and they know that they died out a long time ago. What everyone doesn't know is what killed the dinosaurs. The dinosaurs reigned on Earth from 210 million years ago to 65 million years ago. That's a long reign, historically speaking. Then suddenly, 65 million years ago they died out in a geological heartbeat. The culprit was a natural event. We have good evidence that at this time the Earth was struck by a large meteoroid about five to seven miles in diameter. It struck the Earth just off the coast of the Yucatan Peninsula at what is now Chicxulub. It moved through the atmosphere so fast and struck the ocean with such intensity that it pierced the crust of the Earth and buried itself in the mantle. Coincidentally, many millions of tons of crustal material were thrown into the upper atmosphere and drifted around the Earth distributed by a jet stream. This caused a kind of nuclear winter whereby sunshine was blocked from penetrating the pall, throwing the Earth into darkness. For up to two years photosynthesis stopped. The plants died, as did the herbivores. Carnivores followed suit. About half of the animals on Earth became extinct.

A more recent discovery has revealed there was a second impact of an asteroid about 300,000 years after the Chicxulub impact. Scientists have found a 300 mile wide crater on the Indian Ocean floor which indicates an impact of an asteroid approximately 25 miles in diameter. This impact and its effects would have finished off many of the species of animals that were still recovering from the first impact. About 75 percent of all species of animals were wiped out due to these two events.

The dinosaurs were among the animals that couldn't cope with the new conditions on Earth. Among the animals that survived were types of mammals resembling lemurs and shrews. These animals were versatile enough to adapt and became the ancestors of all modern day mammals.

This mass extinction is not a one-time event. In the past 500 million years there have been at least five major mass extinctions, the largest of which 250 million years ago wiped out 96 percent of marine species and about 70% of plant and vertebrate animal species. These events were so catastrophic that they almost resulted in the sterilization of planet Earth.

You may wonder how we know any of this. Some years ago a team of scientists from Berkeley, headed by Dr. Walter Alvarez, was looking for a scheme of measuring dates. Dr. Alvarez noticed in exposed strata, in a layer about the thickness of a dime, an elevated abundance of rare earths. Elevated in abundance were the elements of iridium, ruthenium, silver, platinum; elements found in abundance in meteorites but not in the crust of the Earth. What could have caused the large quantities of these rare earths?

Alvarez's next step was to find out how widespread the effect was. So he went around the world and dug down to the 65-million-years-ago-level of each place. Each time he found the same thing: an elevation of rare earths. What could have caused a worldwide elevation of materials common in meteorites? The answer was a worldwide infall of meteoric debris at about 65 million years ago—the time of the extinction of the dinosaurs.

Chapter 28

The Sun Also Rises: Or Does It?

Most of the motion we see in the sky is due to the turning of the Earth. This motion is called **diurnal** or daily motion and it causes the Sun to rise and set, and the Moon, planets and stars to do the same. However, although the Sun, Moon, stars and planets *look* as though they're rising and setting, they are not. It is actually the Earth that turns once a day and gives the illusion that these bodies are rising and setting. The axis about which the Earth turns is called the polar axis and it pierces the Earth's surface at two places that we call the North Pole and the South Pole.

Although we know that it is the Earth's motion that causes the apparent motion of the celestial bodies, this is a relatively recent fact, realized for only about 2000 years. The ancients took the Earth at face value; it looked as though everything was moving about us, so they thought that everything was moving about us.

Strangely, astronomers take their cue from the ancients when they discuss the celestial motions. The ancients, believing that we were at the center of the universe and that everything was revolving around us, conceptualized the stars as being attached to the inside of a big sphere. Well, astronomers today do the same thing. They talk of all the stars as *being on* the surface of a sphere of infinite radius, which turns about us once per day. (The sphere turns at a rate of 15 degrees/hour.) It is called the **Celestial Sphere** and we are at its center.

The **Celestial Equator** is a great circle on the celestial sphere. A **great circle** is a circle drawn on a sphere whose center coincides with the center of the sphere.

The **diurnal** or daily paths of the stars are small circles, called **diurnal circles**, whose paths are parallel to one another and perpendicular to the polar axis.

Because the Earth is turning counter-clockwise when viewed from the north, the sky seems to move clockwise. Said another way, when you are facing south, objects, due to diurnal motion, move from left to right, and when you face north, objects move counter-clockwise around the direction of the polar axis.

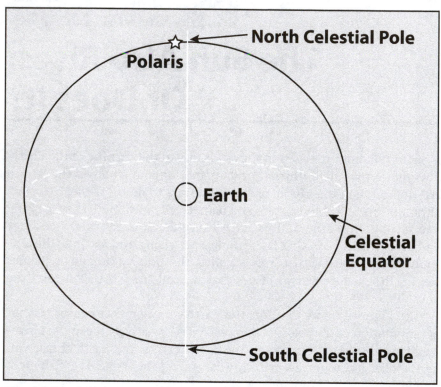

© Kendall Hunt Publishing

The Celestial Sphere

There is a medium bright star near the North Celestial Pole (0.8 degrees away). Because it is so close to the North point on the Celestial Sphere, the star is called "The North Star" or "Polaris." As the Celestial Sphere appears to turn, Polaris actually follows a small circular path, centered on the North Celestial Pole, whose radius is 0.8 degrees.

Because of the geometry of the situation, the altitude of the North Celestial Pole above your northern horizon equals your latitude (distance in degrees from the equator), and because the Pole Star is so near the invisible North Celestial Pole, it serves as a visible marker for it. So, the star is very handy for navigation.

Chapter 29

What's so Blue About a Blue Moon?

We've all heard the expression, "Once in a Blue Moon," and we know this term refers to a small probability. The actual Blue Moon is the second of two full Moons in a single month, which has the probability of occurring once in a Blue Moon.

The Blue Moon is not really blue—it's just an expression. It doesn't look like anything different, just an ordinary full Moon. How often does this happen? On average, there are 41 months with two full Moons per century. That means on average you have one Blue Moon every two and a half years.

What does it mean to have a full Moon anyway? A full Moon occurs when the Moon is diametrically opposite the Sun. If the Moon can't get diametrically opposite the Sun, the full Moon occurs when it is as close as it can get. This occurs every twenty-nine and a half days. So every twenty-nine and a half days there is a full Moon. Now a calendar month lasts, on average, about thirty and a half days, so that means, at best, we have a day to accomplish a second full Moon. So this means in order to have two full Moons in a calendar month, the first must occur on the 1st day of the month and the second on the 31st day of the month.

Chapter 30

"Houston, The Eagle Has Landed": The Epic Flight of Apollo 11

Many of us remember the space race and the Moon landing. Forty plus years ago the United States and the Soviet Union were engaged in a race to the Moon. Just in case you don't remember, or weren't born yet, let me tell you some of the salient features.

On July 16th, 1969, a full six months ahead of schedule (President Kennedy had tasked the nation with completing this goal before the end of the 1960s), the United States launched the mission known as Apollo 11 from Cape Canaveral in Florida (the selection of Florida as launch site came about because it was the farthest south of any launch facility in the continental United States—being south is important because it's where you pick up the sideways speed of the planet as it rotates. The planet rotates a thousand miles an hour at the equator, and zero miles an hour at the pole. Sideways speed is important in getting into orbit, and sideways speed thus gained saves fuel).

The Saturn V (five) rocket, upon which the command module rested, was about 350 feet tall, and is the most powerful rocket to date. This rocket was a multi-stage rocket, so when the first stage had expended itself of fuel, it just dropped into the ocean. At the same time, the second stage fired. When that stage expended itself, it, too, was dropped into the ocean—as the third stage fired. By this time the ship had climbed into orbit.

There were three astronauts in the command module. Mike Collins, who was the command module pilot, was not going to walk on the Moon. His job was to stay with the command module during the descent to Moon's surface. Buzz Aldrin was the one scheduled to be the second man on the Moon—after Neil Armstrong.

The first job in going to the Moon after achieving Earth orbit was to extract the lunar lander from the third stage of the launch vehicle. The service module, on which the command module rested, separated from the third stage, moved off a ways, turned around, and came back and attached itself to the lunar lander, extracting it from the third stage. This was the configuration that was going to fly to the Moon; the lunar module, nose to nose with the command module, and the service module for power, were all that remained of the giant 350 foot rocket that was originally launched. The rockets of the service module were fired, driving the ship out of orbit and on its way to the Moon. The trajectory was a figure eight, around the Moon and back to Earth.

Once the ship got to the Moon, the Moon's gravity pulled it inward. The trick was to go into orbit around the Moon. Energy considerations tell us that if we go to the Moon, the speed will increase so much due to the Moon's gravity that the ship will not stay in orbit around the Moon. Rather, it will fly away from the Moon. But it is trapped in the Earth's gravity, and hence will get a free ticket back to Earth.

This free ticket is what saved a later mission, Apollo 13. The ship suffered an explosion on the way to the Moon, and the men at NASA decided to bring the spacecraft home without landing on the Moon. They simply did nothing when they reached the Moon, allowing the spaceship's velocity to increase as they fell toward the Moon's surface. With the increase in velocity, they did a half orbit around the Moon and then headed back to Earth.

Contrarily, to stay at the Moon, the men of Apollo 11 lost some velocity. At their low approach to the Moon, called Perilune, they fired retro rockets to slow the ship down. This caused them to drop into orbit around the Moon. At this point, Neil Armstrong and Buzz Aldrin climbed through the narrow passageway that separated the command module from the lunar lander, and detached the two ships. The lunar lander, christened "Eagle," moved off to land on the Moon. Using a little retro rocket, the ship descended the few miles to the lunar surface, neatly dodging an impact crater. Aldrin set the lunar lander down.

After an EVA (extra vehicular activity), the men deployed several experiments on the lunar surface and gathered Moon rocks. The Moon rocks were going to tell us about the origin and history of the Moon. The experiments included a seismometer, a lunar laser ranging experiment, and a solar wind experiment. These all took place during the EVA. After the EVA was over, Armstrong and Aldrin climbed into the ship, and using the base of the lunar module for a launch pad, took off in the upper section of the lunar module for a rendevous with Mike

Collins, who had stayed behind in the command module, christened "Columbia."

Now they had to get home. They fired the rockets of the service module to pick up enough speed to break the bonds of the Moon's gravity. Still in the gravitational field of the Earth, they were drawn inexorably closer to Earth. Three days later they arrived. Ejecting the lunar module and the service module, the command module prepared for re-entry. Carefully lining up their orientation to the atmosphere, they prepared to go through it. Too steep an angle and they would be burned up. Too shallow an angle and the spacecraft would skip back off into space.

Successfully completing the transition through the atmosphere where the ship had slowed from 25,000 mph to a mere 200 mph, the astronauts deployed parachutes to slow the remaining speed and drop them safely into the ocean, where they were picked up by helicopter and flown to an aircraft carrier. Since they were the first men on the Moon, they were quarantined for about three weeks because of the threat of a virus, although none appeared.

So ends the flight of Apollo 11.

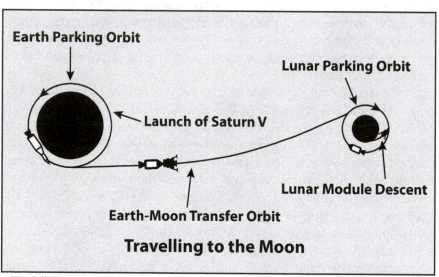

Travelling to the Moon

Chapter 31

A Moon for One Season: The Story of the Harvest Moon

Lore pertaining to the Moon is rife. Particularly abundant are tales of the history of the Moon. For example, the Harvest Moon is said to aid farmers in the field during the months of September and October. What is the Harvest Moon? All full Moons have names. In September the full Moon is called the Harvest Moon. In October the full Moon is called the Hunter's Moon or Blood Moon. The Harvest Moon is the full Moon nearest the Autumnal Equinox. Sometimes it occurs in October, but never more than a week into the month. The Harvest Moon aids farmers by shedding light when light is needed after sunset during the harvest. Normally, the Moon rises about fifty minutes later each day. This means that if the Moon rises at 6 p.m. one night, it will rise at 6:50 p.m. the next night, leaving almost an hour of darkness in between. The next night you will have almost two hours of darkness, and the next night about two and a half.

With the Harvest Moon, position and time conspire to bring the ecliptic parallel to the horizon so that motion along the ecliptic no longer changes the altitude of the Moon relative to the horizon. On successive days the Moon will rise a half hour or less later than the previous day. So for five or six days running, we have a very bright Moon near the time of sunset.

Where and when does this occur? The higher the latitude, the more pronounced the effect. In the northern tier states of the U.S., the Harvest Moon is best observed in September and October.

Chapter 32

A Tale of Dust and Gravity: The Origin of Earth

You may wonder how the Earth formed. It is a tale of dust and gravity full of sound and fury, signifying everything. Picture a giant cloud, five times the size of the current Solar System, about 400 AUs across (an AU, or Astronomical Unit, is the average distance from the Sun to the Earth, or the radius of the Earth's orbit). For billions of years this cloud has been floating stably in space, the gravity being balanced by gas pressure in the cloud. The cloud is neither growing nor shrinking.

Then something happens. We are not sure what it is. Maybe it's radiation pressure; maybe it's a supernova; maybe it's passing through a spiral arm of the galaxy. Whatever it is, it causes a compression in the cloud. When the cloud compresses, the gravitational pull on the surface is increased. The particles that make up the cloud, having been in freefall, are moving faster than they were, accelerating due to the force of gravity. The star-to-be is now unstable, the force of gravity having the upper hand over gas pressure. Smaller and smaller, the star shrinks. The star heats up as it collapses and the center reaches hydrogen fusion temperatures. The star settles down on what we call the main sequence, which is just a phase during which hydrogen is fused into helium. Technically, the star begins its life after the beginning of hydrogen fusion. All the past phases signify the birth of a star and not life per se.

At this point the cloud, which is rotating, will break up into rings of various velocities. Condensations or accretions will form in the bands. As the bands rub on one another, a spin is imparted to the accretions.

The spin leads to the rotation of the planet. The star goes into a phase of stellar winds which cleans out the extraneous matter. The accretions pull themselves together tighter and tighter. These become the planets.

Chapter 33

Fire and Ice: How the Earth Will Die

There's one thing for sure. The Earth will die. But what will kill it? The very thing that gave it life all along—the Sun!

We know, beyond a doubt, that the Sun will die. That's as sure as the day is long. Stars generate energy, and things that generate energy must die in the absence of an infinite energy source. And since there is no such thing as an infinite energy source, the Sun must die . . . and it will take the Earth with it. Here's how it will happen.

The end of the lifetime for a main sequence star like the Sun is reached when the hydrogen is depleted in the core (Where does the hydrogen go? It turns into helium, of course). Then, the reaction which characterized the life of the main sequence star goes out (or at least slows down), for the hydrogen was the fuel for the reaction. The core, which had been held up or stabilized by the radiation pressure, produced by the reaction which turns hydrogen into helium, collapses because the gamma rays, which were the source of the radiation pressure, are no longer being produced. But a gas under compression heats up. So, the core gets hotter. Twenty million . . . thirty million . . . forty million Kelvins. Up and up the temperature goes as the core collapses. Accompanying the core collapse is an expansion of the envelope or outer shell. Why does the outer shell expand? It's because of the hot core.

The core lies next to the envelope. In fact, the only thing different about the core and the envelope is temperature. But now the core is heating up—and it's heating the adjacent envelope material. Now, this adjacent envelope material was just under fusion temperatures when the star was a main sequence star. But now, by being pressed against a heating core, the temperature rises to above hydrogen fusion temperatures. A shell develops within which hydrogen fusion is taking place.

It's called "shell fusion" and it's enough to lift the envelope against the pull of gravity. From the outside, the star looks like it's expanding.

So, when the Sun starts to die, it will start expanding. Over the next ten million years, the Sun will continue to expand. Since expanding gases cool, the Sun will cool as it grows. Its color will start to redden. It will become a red giant star. In becoming a red giant, the Sun will expand between 40 and 200 times (we don't know enough yet to be more precise). But either way, it's enough to fry the Earth. If the Sun gets 40 times as large, it fills the orbit of Mercury, while temperatures on the Earth rise to about 1,200 degrees Fahrenheit. If the Sun gets 200 times as large, it fills the orbit of Earth, and we're toast.

But that's not the worst of it. Within a relatively short time, the Sun will push its outer envelope into space. Such a phenomenon is called a planetary nebula, and once the planetary nebula phase is over, our planet Earth, if it exists at all, will be orbiting as a dead, stellar corpse. This stellar corpse, known as a white dwarf, will have only a small amount of residual energy from the past to warm the Earth, which will freeze. So, from the fires, the Earth will be plunged into frozen darkness, and left to die, not with a bang, but with a whimper.

Chapter 34

Sun Over the Equator: The Autumnal Equinox

September 23rd is a special day. It is the day of the Autumnal Equinox. What does that mean? It means that on that day the Sun traveling southward crosses above the equator and enters the southern skies. It is the mirror image of the Vernal Equinox. Again, as with the Vernal Equinox, the Sun is not moving anywhere. It is the motion of the Earth around the Sun that causes the change. The ecliptic, or path of the Earth around the Sun, extends from +23^1/$_2$ degrees, to −23^1/$_2$ degrees in latitude. When viewed from the perspective of the celestial sphere, the Earth is stationary and the Sun moves about it. The Sun appears to move from +23^1/$_2$ degrees to −23^1/$_2$ degrees in latitude. On March 21st and September 23rd it is crossing the equator. On December 23rd it is at the Tropic of Capricorn, latitude −23^1/$_2$ degrees. On June 21st it is at Summer Solstice, at latitude +23^1/$_2$ degrees. This line of latitude is called the Tropic of Cancer. The reason they are called the Tropic of Cancer and the Tropic of Capricorn is that the Sun was in Cancer and Capricorn on June 21st and December 23rd in the year 150 A.D., when Ptolemy invented the system. So the thing about September 23rd is that it is an equinox, meaning that all over the Earth on that day there are twelve hours of darkness and twelve hours of daylight (equinox literally means "equal nights"). It is one of only two days in the year that this is true—the other day being March 21st.

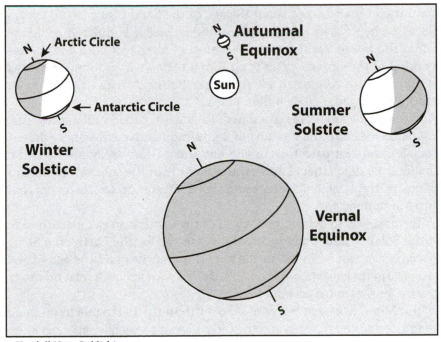

© Kendall Hunt Publishing

Chapter 35

The Age of the Earth

Have you ever wondered about the age of the Earth? It isn't 6,000 years old as Bishop Usher calculated it; nonetheless, it is difficult to determine. The reason for this is that because of plate tectonics and the re-working of the surface, there is no Earth rock we have found that is as old as the Earth. The age of the oldest Earth rock, Acasta Gneisses, from northwest Canada is 4.03 billion years.

The way that rocks are age dated, in general called radiometric dating, is to measure the amount of some radioactive substance (parent nuclide) and compare that to the amount of the decay product (the daughter nuclide); then, knowing the half life of the substance, you can calculate the time since the radioactive substance was incorporated into the sample.

The best way to determine the age of the Earth is to calculate the age of the Solar System and assume the same age for the Earth. The Solar System's age can be found in the age of meteorites and the age of the Moon. The consistency of the results from many different methods gives scientists some certainty.

The Moon has been less worked over than the Earth due to no plate tectonics and weathering, and the oldest lunar rocks have ages between 4.4 and 4.5 billion years. These rocks are the conglomerate gneisses from the lunar highlands.

Meteorites, from radiometric dating, are between 4.53 and 4.58 billion years old. From this, scientists calculate the age of the Solar System to be 4.57 billion years old with an uncertainty of less than 1 percent. All this is consistent, with the age of about 13.2 billion years for the Milky Way galaxy and about 13.7 billion years for the age of the universe.

Chapter 36

Just Who is the Person in the Moon?

The Man in the Moon is a mythical figure, composed of shadowy outlines of several lunar seas, seen from the Earth at or near full Moon. As the "Man in the Moon" is subject to one's imagination, naturally there is more than one candidate outline.

The form of the figure varies with the culture; some of the more prominent ones are these: in America, "Man in the Moon" or "Woman in the Moon;" in Japan, "Rabbit in the Moon." In Nepal, they say that the dead go to the Moon. In Ireland, they say that children come from the Moon. Two of the most prominent forms are shown in the accompanying illustration. The conventional image is a stylized one, in which the left eye (as you look at it) is Mare Imbrium ("The Sea of Showers"), the largest lunar sea; the right eye is Mare Serenitatis ("Sea of Serenity"), the nose is Sinus Aestuum ("Bay of Billows"), and the mouth is Mare Nubium ("Sea of Clouds") and Mare Cognitum ("The Sea That Has Become Known").

The Woman in the Moon is more graphic. Her hair is represented by Mare Serenitatis and Mare Tranquilitatis ("The Sea of Tranquility"). Her eye is represented by Mare Vaporum ("Sea of Vapors"). Her nose is represented by the crater Bode. She wears a pendant represented by the crater Tycho.

The Rabbit in the Moon curls in fetal position, with its head represented by Mare Serenitatis, one ear by Mare Tranquilitatis and Mare Fecunditatis ("Sea of Fertility"), and the other by Mare Serenitatis and Mare Nectarus ("Sea of Nectar"). The body is represented by Mare Imbrium and Oceanus Procellarium.

So we see the Person in the Moon actually has multiple personalities.

© Rafael Pacheco/Shutterstock

Chapter 37

Global Warming

No one anymore seriously doubts that global warming exists (no one, that is, except our government). But despite that fact, there may be aspects of the problem that the reader is unfamiliar with.

There are two types of greenhouse effects—natural and man-made (or anthropogenic). A natural greenhouse effect is caused by water vapor in the atmosphere. This is a good greenhouse effect; it warms the Earth above freezing temperature. Without it, the Earth's temperature would be 30 Celsius degrees colder (54 Fahrenheit degrees). This would make the mean temperature on Earth about 0 degrees Fahrenheit rather than the 59 degrees (as of 2015) it is. The Earth would be a frozen, lifeless ball.

Man-made global warming occurs mainly from the burning of fossil fuels, and mainly it is carbon dioxide that causes the problem. How does carbon dioxide cause global warming? Of the energy that comes from the Sun, some of it causes the heating of the atmosphere. When the atmosphere is heated, it changes the energy into long wavelength infra-red radiation. Some of the radiation is beamed upward and lost to space. However, some of the radiation is beamed downward and trapped in the atmosphere of the Earth. The temperature in the atmosphere rises until the component beamed upward equals the amount absorbed. So the greenhouse effect has a tendency to trap the solar radiation of the Earth.

There are various types of greenhouse gases: water vapor, carbon dioxide, methane, nitrous oxide, ozone, to name a few. Methane is liberated by livestock, and is an important component in the permafrost layer in the Arctic. Nitrous oxide is an important component in fertilizers. Ozone is liberated from car exhaust.

The most foolish thing we're doing, aside from burning fossil fuels, is deforestation. Forests are the means for banking the carbon back into the Earth. Cutting forests down is like cutting out the lungs of the

Earth. We're reducing the forest coverage of the planet by about one football field per second. This loss is taking place preferentially in the tropics where the most biodiversity is extant. We shall pay the price in terms of the rising seas, more tropical diseases, stronger storms, desertification, tropical insect migrations to formerly temperate zones, and loss of biodiversity. If you wonder what you have to do with this, bear in mind that the United States has 4 percent of the world's population but contributes 25 percent of the world's carbon dioxide. We are the problem. When one hundred and fifty nations got together in Tokyo to tackle the problem of greenhouse gases, the United States refused to sign the Kyoto Accord, which would have reduced global warming.

Maybe we should rethink our approach to global warming and the Earth that sustains us?

PART THREE

The Sun, Solar System & Stars

Chapter 38

Solar Energy: Where Does It Come From?

Why does the Sun shine? That question has interested astronomers for many centuries. Back as far as the ancient Greeks, that question was asked and answered. The Greeks thought that the Sun was a hot rock, but it was quickly shown that the hot rock would cool off in a short time compared to the lifetime of the Sun, which at that time was thought to be several thousand years. Another explanation had to be found.

In the 1800s, Lord Kelvin in England hit upon the idea of contraction as the source of energy for the Sun. As the Sun slowly contracted it would liberate gravitational potential energy for a time period much longer than that of the hot rock. But the time period was only ten million years, and geologists were starting to discover the true age of the Earth. That number was in the billions of years. Another source of the energy had to be found.

In the 20th Century, that source was discovered and it was due to Albert Einstein. In the theory of relativity, Einstein suggests that mass can be turned into energy. He made the statement by presenting an equation: $E=mc^2$. This, as it turns out, really explains the energy of the Sun. The Sun is busily fusing hydrogen into helium with the subsequent conversion of mass to energy. In this form of hydrogen fusion, for every gram of hydrogen turned to helium, .007 grams of mass are lost. It has turned into pure energy. This energy is so vast that for every gram of hydrogen turned to helium, enough energy is released to power your home for about twenty years. At this rate you might think that the Sun would burn itself out very quickly. This is not the case—the Sun's mass is immense. With the mass of 2×10^{30} kilograms at a consumption rate of six hundred million tons per second, the solar fuel will last 10 to 12 billion years from the time of origin. Since the Sun is about 5 billion years old, that means it has 5 to 7 billion years remaining.

All the stars in the sky are different ages. Some were born just yesterday in the astronomical scheme of things. The oldest stars in the galaxy are thirteen billion years old. Stars are continually being born and dying. It is the way of the world in the universe.

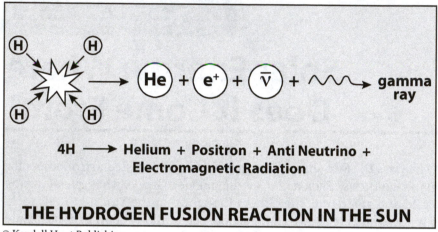

$$4H \longrightarrow \text{Helium} + \text{Positron} + \text{Anti Neutrino} + \text{Electromagnetic Radiation}$$

THE HYDROGEN FUSION REACTION IN THE SUN

Chapter 39

Pluto: The Major Planet That Isn't?

Do you remember when Pluto was a major planet? That changed in the fall of 2006 when the International Astronomical Union demoted the planet. Why did this happen? Pluto was a bad planet. First of all, its orbit is all wrong. The other planets are relatively co-planar, meaning that they all lie in the same plane. Pluto's orbit is inclined by 17 degrees from the plane of the Solar System. Its structure is not planet-like either. It is more like an ice-ball or a comet than it is like a planet. Its size is too small to be taken seriously. Add to all this the fact that Pluto is of very low mass, 1/500th that of the Earth, and you get a picture of a little chip striving to be a planet. The way it was discovered is also odd.

In the early 1900s, Percival Lowell performed a computation to predict where planet X would be. This hypothetical planet affected Neptune and rested outside the orbit of Neptune, which he thought was moving a little bit strangely. He took these perceived strange motions of Neptune and used them to calculate where a hypothetical planet would have to be. He came to the conclusion that there was a planet of six Earth masses out beyond Neptune. When the astronomer Clyde Tombaugh looked for this planet and found Pluto in 1930, it was heralded as Lowell's prediction and discovery. Now it seems it was too hasty. The calculated low mass of Pluto argues that the planet was discovered by accident, for it is nowhere near six Earth masses and the small variations in the orbit of Neptune were not real, only imagined.

You might wonder how it is we can measure the mass of a planet. After all, you can't see the mass. It's not volume, it's not density, it's not weight. The mass of a planet is a measure of the amount of substance the planet has. It stands alone. There is no other property like it. The mass of a planet can only be measured if the planet has a satellite orbiting about it. The orbit of the satellite obeys the laws of gravity and the orbit depends upon the mass of the planet. So the mass of Pluto can be determined if it has a satellite. In 1979, James Christy of the U.S. Naval Observatory discovered a satellite of Pluto, which he named Charon. Due to this satellite, we can calculate the mass of Pluto and find that it is l/500th the mass of the Earth.

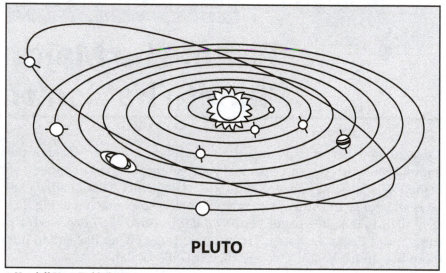

PLUTO

Chapter 40

Comets: A Tale of Two Tails

Everybody knows that a comet has a tail, but do they know that it has two tails?

First, we should define a comet. Comets are ice balls orbiting the Sun in long, elliptical orbits. The typical comet is about one-quarter of a mile in diameter, but large ones can be as much as 50 miles across. A comet is like a large snowball that has been rolled in gravel, dirt and dust: a dirty snowball.

A comet's structure and shape change according to its orbital position relative to the Sun. When it is far from the Sun, it is frozen solid and consists of nothing more than an ice ball with no atmosphere surrounding it. When it is near to the Sun, the Sun's rays cause its icy surface to vaporize, forming a surrounding atmosphere called a coma.

A comet's tail forms only when it is close to the Sun. The Sun's rays push part of the coma back away from the Sun, forming a tail. As well, the pressure from the Sun causes a comet's tail to come in two varieties. The charged particles streaming away from the Sun are solar wind particles, hydrogen ions moving at high speeds. When they interact with gases in a comet's coma, the particles of the coma stream form what is called a gas or an ion tail. Also, there is pressure from the light streaming from the Sun called radiation pressure. This pressure from light acts on dust particles in a comet's coma, pushing the particles away, into a tail called a dust tail. These two tails do not quite point in the same direction although they do generally both point away from the Sun because they are both reacting to the pressure from the Sun. The two tails form a visible "V" shape.

No two comets are the same shape or size and experts can identify a comet from its photograph. A comet's variation, in general, is due to a combination of factors including composition, distance from the Sun, and age. Whether or not a comet has both tails depends on its composition. All extant comets have more or less similar gas or ion tails. Dustier comets have extensive dust tails. Other comets that have made so many orbits around the Sun have no tails because the gas and dust have all been driven from their nucleus. These comets are called defunct comets and they tend to have dirty orbits. They have no nucleus and their dust and gravel particles are distributed throughout their orbit. These dust and gravel particles are responsible for meteor showers.

The words comet and coma are derived from the same Latin word meaning "hair." Comets used to be called hairy stars because their tails flowed behind them, resembling hair streaming in the wind.

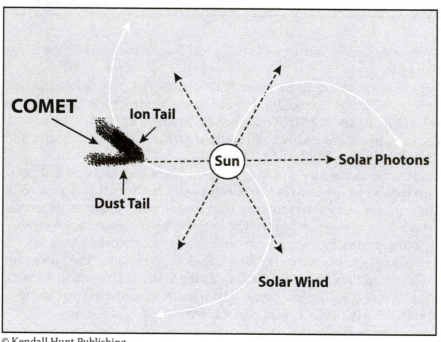

Chapter 41

Ceres: A Not-So-Minor Planet

There is an oddity in the Solar System. The oddity refers to a series of numbers which obey the form $(n + 4) / 10$, where n are the numbers 0, 3, 6, 12, 24, 48 . . . and so on, until you reach the number 768. Now, it so happens that when you put the values for n into the above expression, you generate a series of numbers which happen to be very close to the distances of the planets from the Sun in AUs (one AU, or Astronomical Unit, is the average distance from the Sun's center to the Earth's center). This progression of numbers was discovered by Johann Titius in 1766, and popularized by Johann Bode in 1772. Just what is this Bode-Titius progression? Well, history provides a clue for us. There are two ways to make a discovery in science: theoretically or empirically. Making a discovery theoretically means that there was some other scientific principle which pointed to the new discovery. Making a discovery empirically means doing an experiment and just seeing what happens. When you make a discovery empirically, you generally do not know what is going on at first but, little by little, doing more experiments, you understand the results. You then have theoretical knowledge relating to your discovery.

A case in point for all of this is Kepler's 3rd law. All Kepler could say about his discovery was that he had found that the periods of planets were proportional to their mean distances from the Sun—empirically (by trial and error). But, he had no idea why it was true. The job of understanding Kepler's 3rd law theoretically went to Isaac Newton, who explained it in terms of an artifact of gravity—the why.

When Bode's law was discovered it was not understood. But if there is some scientific reason why planets form where they do, as there must be, Bode's law might be a premonition to that more complete understanding—and much more than a mathematical curiosity.

Whatever the reason, the Bode-Titius progression was taken quite seriously after it was discovered and it was soon noticed that there was a prediction of a body at a distance of 2.8 AUs, a location where nothing was known to exist.

Well, in the best tradition of astronomy, a search was mounted—with great success. A small planet was found on the first night of the 19th Century (January 1st, 1801) by the Sicilian Astronomer Giuseppe Piazzi. But before an orbit could be computed for the body it was lost. Upon hearing about this difficulty, a young mathematician named Frederick Gauss attacked the problem. Gauss had developed a method of orbit calculation based on few variables, the method that is used today to solve problems of orbit determination. Gauss, by November of 1801, had succeeded in calculating the most probable orbit for Piazzi's planet. On December 31st, 1801, Piazzi's planet was reacquired.

The body was named Ceres (after the protecting goddess of Sicily, also known as the goddess of agriculture, and as seen in the word "cereal"). Ceres' average distance from the Sun is 2.77 AUs, in remarkable agreement with Bode's law.

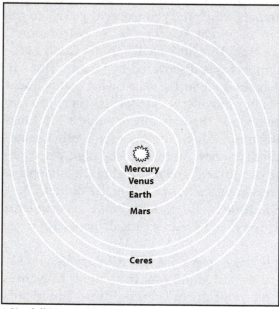

© Kendall Hunt Publishing

Ceres was about 600 miles in diameter, rather paltry for a major planet. After years of study astronomers have found many hundreds of bodies such as Ceres in orbits that average 2.8 AUs from the Sun. This region of space is today called the Asteroid Belt and marks the location of planets that either were torn apart or could never form. Ceres is their chief representative. In 2006, when Pluto was demoted to a "dwarf planet," Ceres was actually promoted to the same status. So, in fact, it is no longer classified as an asteroid.

Chapter 42

Darkness at Noon: Total Solar Eclipses

One of the most rewarding sights in nature is a total solar eclipse. These phenomena are so beautiful that people travel thousands of miles and endure many hardships just to see them. Why is this event called a solar eclipse?

To have a solar eclipse, we need to have a new Moon. In addition, the Moon must pass directly between the Earth and the Sun. When this occurs, the Moon's shadow falls on the Earth and from that spot, we see a total solar eclipse. But total solar eclipses are more than this.

Eclipses are the business of shadows, and it is the shadow of the Moon that is the focus here. Most shadows are composed of two parts: an inner dark part called the umbra from within which you cannot see any part of the light source; and, an outer region called the penumbra, from which you can see part but not the entire light source. If we are on Earth in the circular umbra of the Moon's shadow, which is from 100 to 170 miles in diameter, we experience a total solar eclipse. And if we are in the penumbra, we experience a partial solar eclipse.

As the Moon sweeps around the Earth, the umbra sweeps across the surface of the Earth. The path of this shadow is called the Eclipse Path, or Path of Totality, and people from all over the world try to be in the path of the shadow as it passes by. Because the Moon moves in its orbit around the Earth at 2,000 miles an hour, its shadow moves at the same speed. The Earth is also turning and the speed of its turning depends upon where you are in latitude. For example, at the equator, the Earth turns 1,000 miles per hour and at the spot of each pole, there is no speed at all. Since the shadow of the Moon moves toward the east, the speed of the rotation of the Earth subtracts from the velocity with which the Moon's shadow passes. So the speed of the Moon's shadow relative to an Earth-bound observer is from 1,000 (2,000 minus 1,000) to 2,000 (2,000 minus zero) miles per hour, depending upon where the observer is on Earth.

If you have ever been to a total solar eclipse, you know that these events are accompanied by a whole gamut of effects.

Let's pretend we're at a solar eclipse with a clear sky. The first thing you notice is the Moon placing itself in the way of the Sun. When you look at the Sun (through a filter), there is a little "nick" on one side of the Sun. This is called first contact. Accompanying this effect is a definite cooling of the atmosphere, and winds called "Eclipse Winds." Generally, it's cooler by 20 degrees in the umbral spot than it is in the surrounding area, so air pressure within the umbral spot is reduced. Air from the higher pressure outside of the umbral spot flows into the spot—hence wind.

This intrusion will grow over the next hour as the Moon moves farther and farther into the space between the Earth and the Sun, culminating at the second contact when the Moon entirely covers the Sun. This is when all hell breaks loose. About 30 seconds before totality, the umbral shadow can be seen on the Earth's surface, approaching from the west. It resembles a wall of almost total darkness. Even people who are aware of the nature of eclipses are deeply struck by this. You feel like you want to run away from some approaching doom and you get a notion about why people in the past were terrified of eclipses.

At this point, sometimes another effect occurs—one that is unexplained by modern science. This effect is called shadow bands, which are thousands of bands of light and dark, each about 2 feet across, moving across the landscape at about 20 miles per hour. Shadow bands can be seen before and after totality for about 30 seconds to one minute.

Next we have Bailey's Beads. These are light from the Sun shining through the valleys of the mountains of the Moon. They look like fireworks going off all around the periphery of the Moon. All that is missing is the audible "Pop!" These last a half minute or so.

Then these phenomena play out in reverse order as the Moon retreats. Usually, people are blown away by the experience. You can see pictures of an eclipse and you can be told about the effects of an eclipse, but the only way to really experience an eclipse is to be there.

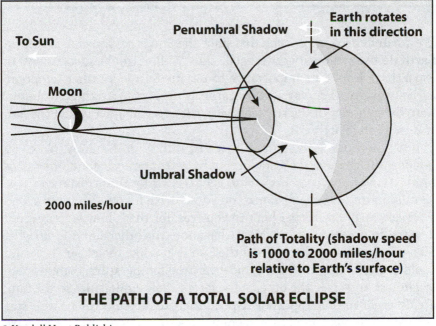

THE PATH OF A TOTAL SOLAR ECLIPSE

Chapter 43

Is the Sun a "Special" Star?

We are under the impression that since the Sun is responsible for life on Earth that somehow it is a "special" star. Well, it might surprise you to learn that, despite its importance to us, the Sun is a rather temporal, garden variety, everyday sort of star. It is no bigger, brighter, or hotter than the majority of the stars in the sky. Its special asset is that the Sun is close to the Earth.

Just how close is the Sun? A mere 93 million miles away. Now you might think this is a far distance, but by astronomical standards this is right around the corner. Let's try to get a feeling for these distances. This 93 million miles is the distance you could drive if you wanted to spend 100 years at 100 miles per hour in your car, for that's how long it would take you to drive to the Sun. Now what about the other stars? From what I have said you would guess that the Sun is just like the other stars, and you'd be right. They have the same measured temperature, composition, brightness and size. The other stars, for all their similarities to the Sun, are different in one important aspect: their distances. The nearest star beyond the Sun is a good case in point. This star, named Alpha Centauri, is 277 thousand times farther from us than the Sun. In other words, 26 trillion miles away. Now this is very far. It would be 100,000 years away in a space ship traveling 25,000 miles an hour. That is the speed at which humans traveled to the Moon.

You can see from the foregoing that travel between the stars is problematic at best. The vast distances involved call for a new technology. For now we will have to wait and look.

Chapter 44

Jupiter: Colossus of the Solar System

It seems fitting that Jupiter was the chief god and the planet Jupiter is the chief planet of our Solar System. But there is no way that the ancients could have known of the planet Jupiter's great size. The planet is eleven times the diameter of Earth, 318 times its mass and three times its gravity. The planet is largely made up of hydrogen, as are three other planets—Saturn, Uranus and Neptune. In the beginning, all of the forming planets were collapsing clouds of mostly hydrogen gas. Some planets retained the hydrogen. Some did not. This was determined by the position of the planet relative to the Sun in the Solar System, and the gravity of the planet. The planets closest to the Sun lost hydrogen whereas planets farthest from the Sun retained hydrogen.

It is obvious how we get the diameter of any planet. We note how big the planet looks and how far away it is. However, determining its mass is slightly more complex. The observed manifestation of mass is gravity. We have found that gravity and mass go hand in hand. The more mass a planet has, generally the more gravity it will have. So measuring mass boils down to measuring gravity. Now how do we get the gravity of a planet? It's simple. We do it by watching a natural or artificial satellite around a planet. The more gravity a planet has, the quicker a satellite will orbit. If we note the time it takes for a satellite to orbit a planet, we can turn that into a measure of the gravity and hence, the mass of a planet.

Although Jupiter is the most massive planet in the Solar System, it is one of the least dense. This is because it is made of a mostly liquid and gas. In fact, Jupiter is a prototype of gas-liquid giants in the Solar System. Calculations of the density of Jupiter show that the planet is liquid hydrogen all the way to a solid core of about twice the size of the Earth. The same is true for Saturn and to a lesser extent, Uranus and Neptune.

Those four planets are known as the gas giants of the Solar System. Together with Mercury, Venus, Earth and Mars—the so-called Terrestrial Planets—they make up the bulk of the material of the bodies that orbit the Sun.

© NASA/Damian Peach

Chapter 45

Mars: Our Next Step in Exploration

Science fiction stories about Mars abound. Why is this so? Well, there is a reason that Mars is the star of so many science fiction movies. Mars has the most Earth-like surface of any planet. Not only are its temperatures hospitable to Earthlings but its distance is not so different from ours. Mars is 150 million miles from the Sun—about half again as far as the Earth. At that distance, the temperatures range from minus 220 degrees to 70 degrees Fahrenheit. The mass of Mars is $1/10^{th}$ and its diameter is 1/2 of Earth's, which lead to a gravity of 40 percent of Earth's. Due to this low gravity, the pressure of the atmosphere is $1/100^{th}$ that of Earth. At this pressure, liquid water cannot exist. The atmosphere is 96 percent carbon dioxide. Relative to other planets, all these numbers are not so bad. We can work there by wearing heated pressure suits and oxygen masks.

Mars is graced with north and south polar caps, which wax and wane with the seasons.

Two difficulties with Mars are that there is no liquid water and that there is no oxygen in the atmosphere. But in fact, there is evidence that liquid water once did exist on Mars. Originally, Mariner 9, in 1970–1972, took photographs of Mars, which showed clear evidence of channels made by running water. More recently, spacecraft and rovers that we have sent to Mars have found great evidence for ancient Martian riverbeds, oceans and flooding. In fact, there is strong evidence that water still flows under the surface of Mars. Obviously Mars, in the past, had a different climate than it does now. There must have been a thicker atmosphere in the past as well. The atmosphere is now locked in the polar caps, which are made of frozen water and frozen carbon dioxide. So if we go to Mars, it would be our job to liberate the atmosphere. One way to do this is to cover the caps with something dark. That something could be plants. Dark polar caps would absorb more heat from the Sun and this would melt them. The plants would then absorb the carbon dioxide and liberate the oxygen and water. The atmosphere would thicken and liquid water would be able to exist once more on the surface.

It is likely that if we ever do colonize another planet, that planet will be Mars. All the other planets are far too inhospitable. Of all the planets, Mars is the only one whose conditions are at all manageable for eventual human habitation.

Chapter 46

Mercury: Now You See It, Now You Don't

The planet Mercury is the swiftest planet of them all, charging around the Sun in 88 Earth days. Its speed is so great that its motion from night to night is quite apparent. It is strange to say, however, that few people have seen this planet even though it is quite bright. The reason for this is that Mercury never gets more than 28 degrees from the Sun (from an Earth-based observer's viewpoint), a product of its small orbit, so it is always seen hanging very close to the Sun in the sky and therefore difficult to detect. The only time you can see the planet is just after sunset or just before sunrise when it is at its so-called greatest elongation. The elongation angle is the angle between the Sun and Mercury as seen from the Earth. The planet itself is 3,000 miles in diameter, or 3/8ths the size of the Earth. It is 1/20th the mass of the Earth and has no atmosphere to speak of because of its weak gravity and high temperature. It has a highly eccentric orbit, its closest distance to the Sun being about 29 million miles and its furthest being about 43 million miles.

The planet rotates in a period of 59 days, exactly 2/3rds of its orbital period of 88 days. This means that each time the planet is at its perihelion point (closest to the Sun in its orbit), it turns one face or its antipode toward the Sun. This is a result of having two tidal bulges on the planet which are alternately facing the Sun when the planet is at close approach with the Sun.

Mercury mythologically was the god of speed and the messenger of the gods. He was Hermes to the Greeks, and Woden, god of the sky, to the early Germans.

Chapter 47

Minor Planets: Are They a Threat to Earth?

In the space between Mars and Jupiter is the Asteroid Belt or zone of minor planets. One of these bodies, named Apollo, is the prototype for a family of bodies called *Apollo Asteroids*, which are of particular interest to terrestrial astronomers because they have Earth-crossing orbits. Bodies with Earth-crossing orbits stand at least a chance of colliding with the Earth. Since these asteroids can be up to 10 miles in diameter, they can be very dangerous.

Consider a case in point. The historical record of the Earth clearly demonstrates the fact that there have been several "mass extinctions" presumably caused by collisions with Apollo asteroids. Such an event might look like this:

It had been floating in the vacuum of space for 5 billion years. At a diameter of five miles, it was about average size for its type. It had been in this vicinity before, countless times. Only this time, there was a planet in the way. It started with a gentle gravitational tug that built and built until the collision was a mathematical certainty. The rock went through the atmosphere as though it were not there, traversing it in 5 seconds. It hit the ocean, which it recorded as "some dampness." Then it exploded through the crust and buried itself deep within the mantle. Billions upon billions of tons of crustal material were thrown into the upper atmosphere where the particles were caught in the jet stream, carried around the Earth, and held aloft for as long as two years.

Here's the point: Because the sunlight was effectively blocked from hitting the planet's surface by the dust for as long as two years, there was a cessation of photosynthesis which first wiped out the plant eaters, which in turn wiped out the carnivores that ate them, and the carnivores that ate the secondary carnivores, and so forth up the food chain.

All in all, it was a pretty good day for the mammals who were versatile enough to survive during the two years following the impact and could take advantage of the gap left by the dinosaurs and come to rule planet Earth.

Chapter 48

Saturn: Rings, Rings, and More Rings

The jewel of the Solar System is Saturn. Its bright rings can be seen for billions of miles. The planet is actually a gas giant, like Jupiter, and second to it in diameter. It is a planet that is liquid almost completely to the center. Its atmosphere is a swirling mixture of hydrogen, helium and methane gases, which are in constant turbulent motion. One unique property of Saturn is its density. It is the only planet in the Solar System with a density less than one gram per cubic centimeter. This means that if you had a bathtub large enough and you put Saturn in it, it would float (but it would probably leave a ring!).

The thing that makes Saturn special is the rings. We now know from the Voyager and Cassini probes that the rings are composed of chunks of ice, all floating in their own separate orbits around Saturn. From the Earth we can only see three broad rings, which we have classified with the poetry of the scientists, the A Ring, the B Ring, and the C Ring. But upon closer examination, the rings resolve themselves into many thousands of ringlets. The rings of Saturn look like a giant phonograph record. What keeps them all in their places? The theory is that the gravity of some of Saturn's satellites, called shepherd satellites, coaxes the ice boulders into their various orbits and keeps them there. It turns out that not only Saturn has rings—all the gas giants in the Solar System do. The rings of Jupiter, Uranus, and Neptune are much fainter than the rings of Saturn and we only know about them because of the space probes we have sent.

As of 2016, sixty-two satellites have been detected orbiting Saturn, the number being based on their size. The largest satellite is Titan, the only satellite in the Solar System to have an extensive atmosphere. Our measurements show that the atmosphere of Titan is mostly nitrogen at a pressure of 1.6 times Earth's atmospheric pressure. Also present on

Titan are methane and ammonia. In fact, methane acts as water on Earth does, existing in gaseous, liquid, and solid states. So on Titan we can have lakes of liquid methane with frozen methane floating in them as icebergs. The atmosphere of Titan is an opaque mixture of organic material rendering its surface invisible to us. However, in 2006, we landed a spacecraft, called the Huygens probe, on Titan. It was jettisoned from the Cassini spacecraft which was orbiting Saturn. It transmitted data and sent photographs for over an hour. It was the first time one of our spacecraft landed on the satellite of another planet.

In Greek mythology, Saturn is the father of Jupiter, and is famous for eating his children. Saturn is also known as Cronus, god of time.

Chapter 49

The Day the Music Died: The Death of the Sun

It seems hard to believe the Sun will not last forever. There will come a day when the Sun will die, not with a bang, but with a whimper; not quickly, but slowly. The story of the death of the Sun is a very serious matter because the odds are that some day people will have to face it.

The Sun is a nuclear furnace converting hydrogen to helium in its core. If hydrogen is the fuel of the Sun, the Sun will die when the hydrogen is used up. The sum total of hydrogen allotted to the Sun never increases. It always diminishes at a rate of 600 million tons per second. This is the rate that hydrogen turns into helium. When the hydrogen is all depleted—when there is no more—the reaction that powers the Sun, called "the proton-proton reaction," will slowly fade away. The reaction will fade away over the next 5-7 billion years. During that time the core will slowly shrink because the gamma rays which have been the hallmark of the proton-proton reaction will disappear. Those gamma rays had been pushing outward toward the surface of the Sun and providing the outward pressure to fight off gravity. Now with the gamma rays gone, the core of the Sun collapses and in response the envelope or outer shell of the Sun will start expanding. This is because the core, when it collapses, heats up and initiates a region of shell fusion. Shell fusion radiates so much energy that the envelope is pushed outward.

The net effect of all of this is that the star collapses in the core while the surface is expanding. Now we know that expanding gases cool, so the star on the outside cools down and evolves to become a red giant. The process just described takes about 10 million years. The star remains a red giant for up to 1 billion years. This is about 10 percent of its lifetime, so we can see that the Sun will slowly grow and die in place in the Solar System. The planets will continue to orbit the dying star.

The outer shell of the Sun will continue to expand and be pushed slowly outward into space, ejected like a soap bubble. We see stars like this all the time. They are called *planetary nebulae.* Not because they have anything to do with planets, but because when they were first discovered the blue-green disk-like appearance reminded astronomers of the planets Uranus and Neptune. An example of a planetary nebula is the Ring Nebula in Lyra. Over a time period of 20–30 thousand years the Sun, having expelled its envelope, will just be a bare core—the subject of my essay, "The Expansion of the Universe."

Chapter 50

The Return of Sol: The Winter Solstice

There is a common misconception that on December 21st or 22nd the Sun stands still. Even the name Solstice bespeaks a lie. Solstice means stationary Sun. Actually, what is happening is that the Sun is ceasing its southward motion and then starting its northward motion. If the Sun was going south and now it turns northward, doesn't it have to stop first? No, you see the Sun is moving eastward, first southeast and then northeast, it never stops its eastward motion, just its southward motion (see diagram).

What is this motion, anyway? It is the Earth's revolution around the Sun. This makes it appear as though the Sun is moving around us, but actually we are moving around the Sun. This motion is continuous; we never stop and back up. The only thing that happens is that the Earth, tipped on its axis, reaches a point in its orbit where its northern axis is tilted in a direction away from the Sun.

How fast does it move? The Earth moves 360 degrees around the Sun every 365 days. That is a little less than a degree per day. Actually the ancient Egyptians had a false impression of the length of the year. They thought it was 360 days. They were wrong. The Greek astronomer Hipparchus was the first to measure the length of the year accurately.

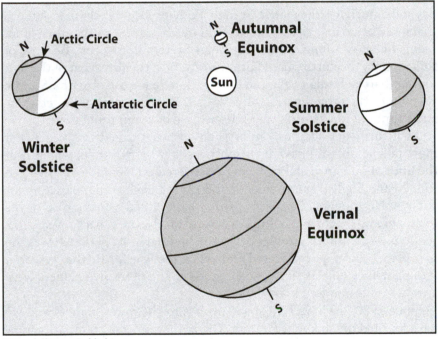

© Kendall Hunt Publishing

Chapter 51

Uranus: The Tipped Planet

Beyond Saturn lies the planet Uranus. Perhaps its chief claim to fame is that its rotation axis tips to the vertical by 98 degrees so that the planet seems like it is rolling around the Solar System in its orbit. Because of this motion, the north pole of Uranus points in the direction of the Sun for one-half a Uranian year and the south pole for the other half of the year. Since the orbital period of Uranus is 84 years, each pole gets constant sunlight for 42 years, an extreme case of the midnight Sun.

As it turns out, Uranus can be seen with the naked eye, a fact that most people do not know. Its brightness is 6th magnitude, which is at the limit of human visibility. It was discovered accidentally one night in 1781 by the English astronomer, William Herschel, while searching for comets. He thought the object, which looked like a fuzzy disc in his telescope, was a comet. But upon careful study, the object revealed itself to have an almost circular orbit, the hallmark of a planet. It was the first planet that was discovered and paved the way for further research into planetary objects in the Solar System. The other five planets were always visible and well known.

Since 1980, we have sent space probes to the outer planets and we have learned much about Uranus. The planet is a gas giant, as is Jupiter. It is bluish in color, the product of methane in its atmosphere. It has a seventeen hour day and it is four times the diameter of Earth. The planet has about a dozen narrow rings which are very dark, the product of carbon soot. There are twenty-seven satellites out of which five were discovered from Earth. They are: Ariel, Umbriel, Titania, Miranda, and Oberon—all named for sprites and faeries from English literature.

A unique body in the Solar System is Miranda, a satellite which looks as though it were made out of several diverse types of material, all mixed together. Its surface is a jumbled combination of blocks of ice and rock. Its unique feature is a cliff nine miles high, which is nine times higher than the top of Half Dome to the valley floor in Yosemite National Park.

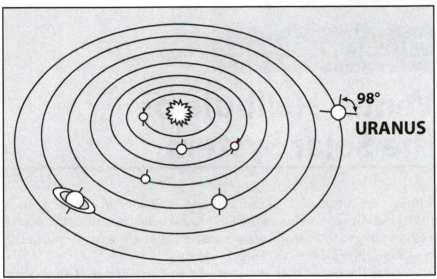

Chapter 52

Venus: Hellhole of the Solar System

Named after the goddess of love and beauty, Venus (known as Aphrodite to the Greeks and Isis to the Egyptians) belies this title. The planet that is so bright and conspicuous in our night sky is actually one of the most dastardly places in the Solar System.

Venus is the planet that orbits nearest to Earth and is of almost the same size and mass as Earth. For that reason, it is sometimes called Earth's twin. But the similarity stops here. Discoveries of the conditions on the surface of Venus show it to be the most inhospitable planet in the Solar System.

The brightness of the planet is because it is close to us and is cloud covered. At first, the clouds were taken to mean that the planet was moist, but more recent observations show that the clouds, rather than being made of water, are a concentrated solution of sulfuric acid. What is more, the measured temperature of the planet, rather than being a piping hot 140 degrees Fahrenheit as it should be, given its distance from the Sun, is an astounding 900 degrees Fahrenheit. This discrepancy in temperature is due to a runaway greenhouse effect caused by an almost total carbon dioxide atmosphere.

If this were not enough, the atmospheric pressure of Venus is extreme. Rather than the pressure of 15 pounds per square inch, as it is on Earth, the pressure on Venus is 1,300 pounds per square inch. The reason for this is that the mass of the carbon dioxide atmosphere of Venus is some 90 times the mass of Earth's atmosphere. This is so because at its early stages of evolution, Earth dissolved carbon dioxide in its oceans, which were carbonated soda pop. Eventually, the carbon precipitated out in the form of calcium carbonate, which became the limestone of the sea floor. At the temperature of Venus, water cannot condense into oceans. There is no place for the carbon dioxide to go, so it remains in Venus' atmosphere. If we want to the Earth to start resembling Venus, we should continue to pour carbon dioxide into our atmosphere.

Chapter 53

Neptune: The Planet that Refused to be Discovered

When William Herschel discovered the planet Uranus in 1781, it caused quite a stir because nobody had ever discovered a planet before. But then, if there was one new planet out there, why not two or even three? The search was on.

By 1790, a precise orbit was calculated by astronomers for Uranus, taking into account all of the then-known gravitational forces acting on the planet. Much to their consternation, the computed motion of Uranus could not be reconciled with its past motion. The two were different by as much as 2 minutes of arc (1/30 of one degree). This might not seem like much of an error, just barely visible to the human eye, but in celestial mechanics, the science of movement in the universe, it's a monstrous error and must be accounted for.

In 1841, a young Cambridge mathematics student, John Couch Adams, had a remarkable idea. What if the difference between the observed and predicted motion of Uranus were due to the gravity of an invisible planet lying somewhere beyond Uranus? To test the idea, Adams used Newton's laws to compute just where such a planet would have to be to cause the observed differences (called *residuals*). When he arrived at the answer, in October 1845, Adams sent a letter to the Royal Astronomer, Sir George Airy, informing him of his theory and of his solution. But Airy was a very busy man and he had no faith in this unheard-of astronomer, and he promptly forgot the whole thing.

Meanwhile, in France, the astronomer Urbain Le Verrier had a similar idea to Adams' and he arrived at a similar answer (only different by 1 degree). He published his results in June, 1846. The similarity of the predictions of Adams and Le Verrier caught the attention of Airy, who finally decided that someone should take a look at the whole affair. So, he contacted J. Challis, the Director of the Cambridge Observatory, and he asked him to take a look at the Aquarius region of the sky, where the planet was predicted to be. But Cambridge, as it happened, had no up-to-date star charts of the Aquarius region. In the absence of these charts, Challis had only one recourse. He had to sketch the region (today, we would use photography), marking down all the stars, wait several weeks and then do it again. Since planets orbit the Sun, the positions of the planets change, i.e., the planets move. Challis dutifully made his first observations and unknown to him, he sketched the planet Adams and Le Verrier were looking for. Unknowing, he put them in a drawer, meaning to go back to the project. But he never did go back to the task and the planet remained undiscovered.

Finally, Le Verrier took the bull by the horns. In 1846, he contacted Johann Friedrich Galle at the Berlin Observatory. He knew that Berlin had up-to-date star charts, which would aid in finding a hypothetical planet. Galle received Le Verrier's letter on September 23, 1846. That very night he went to the observatory and he found the planet within 20 minutes of the start of observations! It was within 1° of Le Verrier's prediction.

The discovery of Neptune was hailed as a triumph for, and a confirmation of, Newton's laws. After all, the laws correctly predicted the whereabouts of a planet. Neptune was the first planet discovered with a pencil and paper.

Chapter 54

Driving to Pluto: The Size and Scale of the Solar System

I've often thought that a good way of getting students to internalize the size of the Solar System is to talk about it in terms of a motor car and driving. Students seem to naturally identify with the automobile, so I offer this scenario: let's say there are roads throughout the Solar System; long, smooth, and straight. Further, let's say you decide to drive on these roads at 100 miles an hour, a fair speed. How long would it take you to get to the Moon? We know the Moon is, on average, 240,000 miles away. Traveling at 100 miles per hour and not stopping until you get there, you could drive 2,400 miles a day. Thus, it would take you 100 days or three months to arrive at the Moon. This isn't an outrageous idea. I'll bet if there were such a road you'd find cars on it because three months is less than the time it took California-bound 49ers to cross our country. There's always somebody looking for adventure.

But now all bets are off because the next object of interest is the Sun. How long would it take to drive to the Sun, the radius of our orbit? Well, the Sun is 400 times further than the Moon. So it would take 400 times a quarter of a year or 100 years to reach the Sun. Imagine sitting behind the wheel of your car for 100 years.

Now what about the distance to Pluto? The Earth is one-fortieth of the way to Pluto from the Sun. So that means to drive to Pluto would take 4,000 years. It's mind-boggling. Even in our little Solar System the distances are vast. You may wonder how it is that we get space probes to go to the outer Solar System in a reasonable time. The answer is we don't go at 100 miles per hour. A recent space probe to Pluto traveled at 36,000 miles per hour and took 9 years to get there. Another recent space probe that went to Jupiter traveled at 164,000 miles per hour and only took 5 years to get there. So the speed of the ship is very important. Let's take one practical example: the speed of a Moon ship is 25,000 miles per hour. How long would it take to get from Earth to the nearest star at that speed? By the way, the nearest star is 4.3 light years away or 25 trillion miles. To make the trip would take you over 100,000 years at 25,000 miles per hour. So you see, even at fast speeds, star travel is out of the question for us at this time.

Chapter 55

Orbits: How They Work

There's nothing difficult about orbits. Orbits are simple things. You have small bodies and large ones and typically the small body goes around the larger one. Let's see how it happens. We find with orbits that an impulse is needed to get the object moving, such as from a rocket. Once we have that, then it's just business as usual. We have an orbit.

Isaac Newton, King of Orbits, explained it this way. Say you have a tall mountain that extends above the atmosphere. You go to the mountain, taking with you several balls. Once at the top you drop a ball. It falls straight down and lands on the mountain's flanks at a point directly beneath the release point. The next thing you do is throw a ball sideways like skipping a stone. What will happen? The ball will drop and move sideways at the same time. It will move in an arc. Next, throw another ball sideways even harder. The ball will move farther from the mountain before it hits the ground.

Keep doing this, harder every time, and the trajectory of the balls will open. There is one speed for which the trajectory will lie parallel to the curve of the Earth. It is called the circular satellite speed and it means that the ball gets no closer to the Earth as it falls. It is in orbit. The only thing that can modify this situation is if the ball is in the atmosphere. It will then slow down due to air resistance and the orbit will decay. That's why you have to send the rocket up to put it in orbit. It has to be above the Earth's atmosphere. If it were not for that reason we could put satellites in orbit at any height above the ground. For example, if the Earth were devoid of any atmosphere you could shoot a projectile horizontally at a height of, say, 5 feet at 5 miles per second toward the east and then 90 minutes later you could look for it coming over the western horizon, speed undiminished. You would have put a body in orbit 5 feet above the Earth.

This scenario can be practiced on the Moon which has no atmosphere. On our satellite you can put bodies in orbit at any height above the surface.

So the answer to the question, "How many forces does it take to keep a body in orbit?" is one, which is "gravity."

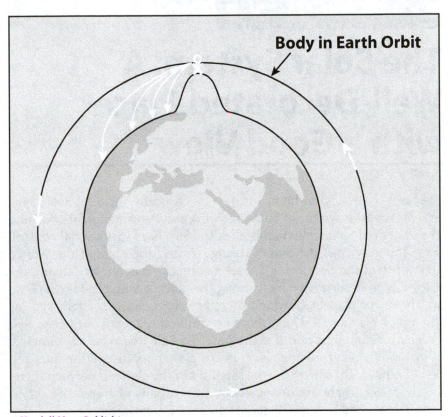

Body in Earth Orbit

Chapter 56

The Solar System: A Well-Decorated Place with a Good View

The Solar System is our home in space. Technically, the term Solar System means "the system of Sol," Sol being the Latin name for the Sun. The system of Sol includes all bodies the Sun has gravitational control over. The gravity of the Sun is strong because the mass of the Sun is great. In fact the Sun, being a star, has much more gravity than anything else in its vicinity. Chief among the objects controlled by the Sun are the major planets, of which there are eight. These are followed by the minor planets, of which there are millions—mostly in a zone between Mars and Jupiter, called the Asteroid Belt. These are the remains of material which could not form into a planet because of the tidal action of Jupiter. Then there are the satellites of the planets, of which there are over 150. There are also comets, frozen balls of water and other compounds, and at large distances from the Sun in places called the Kuiper Belt, which starts just beyond Neptune's orbit, and the Oort Cloud, which may lie up to 1 light year away from the Sun. Lastly, we now have bodies designated as dwarf planets. These bodies are spherical, like major planets, but have not gravitationally cleared their orbit. In other words, they orbit around the Sun with other bodies very close to their orbit. Pluto is now classified as a dwarf planet.

The Solar System is a highly organized place. All major planets and their satellites are on roughly the same plane, along with most of the minor planets. All planets move in the same direction around the Sun and most rotate in the same direction. The inner planets are small rocky worlds, while the outer planets are large gas-liquid giants. The spacing of the planets is similarly regular. The inner ones align closer together, while the outer ones spread farther apart. All this regularity calls for an organized formation for the Solar System. The theory is that star systems (systems like our Solar System) are formed when a giant cloud begins to collapse. This so-called pre-solar nebula starts collapsing because of a relatively nearby supernova explosion, or perhaps from passage through a density wave in the galaxy. Because of conservation of angular momentum, the spin of the cloud speeds up, forcing the cloud to flatten out into a form like a pizza (imagine a pizza being made from somebody tossing up and spinning a ball of dough). Within the spinning cloud, sub-concentrations of matter form due to gravity. These are the planets.

This scenario neatly explains the observed properties of our Solar System. All the planets are in the same plane because of conservation of angular momentum. All the planets rotate in the same direction because that's the way the cloud rotates. The inner planets are small, so there can be smaller spaces between them, while the outer planets are large, so they need larger spaces. This hypothesis neatly explains the origins of the system and makes it clear that it can happen elsewhere.

Chapter 57

Planet X: Does It Exist?

What is Planet X? Planet X is the name given to any unknown and undiscovered planet beyond the orbit of Uranus. Throughout the course of history there have been many predictions by many people of some Planet X. The most famous of these was Lowell's prediction of a Planet X which turned out to be the infamous minor planet called Pluto. Pluto's demotion into dwarf planethood was big news in 2006. What happened was that a spherical object which was a little larger than Pluto was discovered in 2005. This object was given the catalog name 2003 UB313, whimsically dubbed "Xena", and then later, officially called Eris. It would have been the 10th major planet and astronomers were thinking that many more major planets would be discovered after Eris. Therefore, the International Astronomical Union decided to classify Pluto, and any other body like Pluto, a "dwarf planet."

This is likely the way it's going to be from now on. We may discover more planets out there in the Kuiper Belt or maybe even the Oort Cloud but they will not have the stature of major planets. They will be more likely to be snowballed like Pluto or Charon than they will be like Jupiter. Planets seem to increase in percentage of water with distance from the Sun. Near the Sun, the volatile water evaporates readily, so the planets near the Sun have little water. As a planet gets farther from the Sun, the percentage of water increases dramatically, until by the time you get to Neptune, the planet can be half water. So most of the dwarf planets that have now been discovered are bodies composed of ice and rock. We have confirmed 5 dwarf planets so far but feel there may be up to 200 in the Kuiper Belt and thousands within the confines of the Solar System. So even though we may not have a massive "Planet X" lurking somewhere out there in the Solar System, we know we have many little Planet X's milling about.

Chapter 58

A Mysterious Wind

There is a mysterious wind in the inner Solar System: the solar wind. To understand the solar wind, you must first understand the regions of the Sun. In simple terms, the Sun is divided into four regions. The first region is the only one we can never see: the interior. This region contains the core, and the envelope. The envelope is a region of convection which carries energy from the core to the photosphere, which is the second region. It is the region that you see when you look at the Sun. It is 150 miles deep and about 11,000 degrees Fahrenheit. Within it some major magnetic phenomena take place, the most important of which are sunspots.

The third region is the chromosphere, on top of the photosphere, so-called because of the red color of its "flash" spectrum. Astronomers have had a difficult time observing just the chromosphere. This is because its spectrum (called the "flash" spectrum) can only be seen for a few minutes before and after totality during an eclipse.

But we're really interested in the fourth region, the solar corona (crown): the two million degrees Kelvin region surrounding the Sun, and the aim of every eclipse seeker. There is something very strange about the corona. What we don't understand is the cause of its high temperature. The photosphere of the star is cooler than its interior, as it should be, since the photosphere is gaining its power from the interior (11,000 degrees Fahrenheit for the photosphere, compared to 27 million degrees Fahrenheit for the core). But here the temperature reverses and goes up with distance from the Sun. It shouldn't do that under ordinary circumstances. One recent theory is that there are "tornadoes" of hot gas reaching up into the corona due to intensified swirls of the Sun's magnetic field. This would lead to a rising of temperature. At the corona another strange phenomenon exists: it is expanding. The farther you get out in the corona, the more rapidly it is expanding away from the Sun. At one solar radius, the corona is expanding at one hundred kilometers per second. At two solar radii, the corona expands at two hundred kilometers per second. At the Earth's distance from the Sun, the gas has accelerated to about four hundred and fifty kilometers per second.

The mass loss the Sun sustains is prodigious from an Earthly point of view. About 4 billion kilograms (8.8 billion pounds) per second are lost. The material is 95 percent singly ionized hydrogen (protons) and 4 percent doubly ionized helium atoms (called alpha particles).

We see then that the temperature has been turned into kinetic energy. Where the planets live, this coronal gas is called the solar wind. Since we are at the outer extent of the corona, we are in the solar wind. So are Mars, Jupiter, Saturn, Uranus, Neptune, and Pluto. The solar wind extends at least twice as far as Pluto before it merges with the general magnetic field of the galaxy, at which point it is called the heliopause.

There are several effects for which the solar wind is responsible. On the positive side, a comet's gas tail is pushed back from the comet by the solar wind. The tail always points away from the Sun, evidence that it is being pushed by the solar wind. The aurorae are a result of solar wind particles hitting the atmosphere. On the negative side, magnetic storms, caused by solar wind, create communication disturbances and may present a hazard to astronauts in space.

Chapter 59

Spots on the Sun: The Sun's Magnetic Cycle

The Sun is like a giant magnet in space. Geographic poles are superimposed on its magnetic poles, one north, one south. Now, the Sun's magnetic field does something very strange. Every eleven years the poles change polarity; north becomes south and south becomes north. The whole general field of the Sun reverses (the Earth does this, but only after several thousand years).

Now, this is not the only major magnetic property of the Sun. There are dark regions on the Sun—called sunspots. In certain instances, large ones can be seen with the unaided eye, especially if the Sun is low on the horizon or viewed through a layer of obscuring clouds or fog. Because Aristotle praised the perfection of the heavens, people wanted not to believe these imperfect sunspots could be on the Sun. Rather, they thought what they were seeing were silhouettes of opaque vapors floating in the Earth's atmosphere and projected onto the Sun.

In reality, sunspots are magnetic storms in the photosphere of the Sun. Their dark appearance indicates merely that they are cooler than the surrounding photosphere, about 2,000°–3,000° cooler (the photosphere is 11,000° F, while the sunspot is between 8,000° and 9,000° F).

Sunspots last from hours to months, and are made up of a dark central region (the "umbra") and a lighter gray perimeter (the "penumbra"). They frequently occur in groups of two or more. Since the Sun is rotating, the spots in a sunspot pair are designated as the preceding (P) spot and the following (F) spot. These sunspots are highly magnetic and the P and F spots in a typical group are generally found to have opposite polarity (one spot will have north magnetic polarity while the other spot has south magnetic polarity). The polarity of the P spot will always be the same as the general polarity in the hemisphere in which the spots lie.

In the middle of the 19ᵗʰ Century, it was discovered by the German amateur astronomer Heinrich Schwabe that the number of sunspots visible in the solar photosphere varied with time. Sometimes many spots were visible while at other times none were seen. A study showed that the period of the variation was between 8 and 18 years with an average of about 11 years.

At the beginning of a new cycle, which occurs just after "sunspot minimum," a few spots appear at mid-solar latitudes (about +30 and −30 degrees). As the cycle progresses, new spots appear at lower solar latitudes until "sunspot maximum," when spots appear near latitude +/−15 degrees. Near sunspot minimum, the few spots that appear are found near solar latitude (+/- 7 degrees).

At sunspot minimum, the overall general magnetic field of the Sun fades and when the new cycle begins (with a few spots appearing at high latitude), the general polarity is reversed along with the polarities of the P and F spots relative to the preceding cycle.

Chapter 60

A Cosmic Year: The Solar Odyssey

There are many strange units of measurement in astronomy: AUs, solar masses, and parsecs, just to mention three. Yet another unit that is wholly obscure is found in galactic astronomy: the cosmic year. It is defined by analogy to the Solar System, where the Earth is defined as taking one year to orbit the Sun, as the time necessary for the Sun to pass once around or orbit the galaxy. It is a unit of time that is equal to 200,000,000 to 250,000,000 years. In units of cosmic years, the universe is 60–75 years old, and the Solar System is 20-25 years old. Translated into Earth terms, one cosmic year ago the Earth was just entering the age of the dinosaurs, which ended about 1/4 of a cosmic year ago.

The period of time that is one cosmic year is calculated from knowledge of the radius of the orbit of the Sun around the galactic plane and the velocity of the Sun in the galactic plane. If we divide the distance traveled by the Sun in one orbit (from the radius of the orbit) by velocity of the Sun, we get the number of miles in a cosmic year; 2.5 quintillion miles or 2.5×10^{18} miles.

PART FOUR

The Milky Way, Galaxies & Cosmology

Chapter 61

Are They Out There? The Discovery of Exoplanets

Throughout the latter half of the 1900s, there were two main trains of thought regarding extraterrestrial civilizations. First, there were the UFO (Unidentified Flying Objects) enthusiasts who felt that the Earth has been, and continues to be, visited by beings from other planets beyond the Solar System. Then there were the UFO skeptics, who, despite thinking that aliens may be out there, felt there was never good enough evidence to demonstrate any visitation at any point. It seemed to the skeptics that as Carl Sagan quoted, "Extraordinary claims require extraordinary evidence," that was exactly where we were with the claim that we have been visited. There was no compelling evidence. Despite one's belief—visitation of extraterrestrials, or not—there was an underlying question that emerged. That is, "Are they out there?" To answer this question, we had to know if there were actually other planets beyond the Solar System and if so, could any of them support life as we know it. Before the 1990s, nobody knew.

Then in 1995, everything changed. A team of astronomers discovered the first exoplanet (or extrasolar planet) orbiting around a main sequence Sun-like star* in the constellation of Pegasus, the winged horse. The planet was called 51 Pegasi b and was a gas giant planet similar to Jupiter and Saturn. This planet was not discovered by "seeing" it but was detected by the wobble of its parent star as the planet orbited it (the Doppler Effect was utilized to determine slight variations in the star's spectrum due to this wobble). This was a monumental discovery, as the world finally knew that we were not alone—at least there were other planets out there. However, did that mean other life? Not necessarily. Many other planets were discovered very shortly after the first and most had the same characteristics, very massive gas giants and very close to their parent stars. Even though discovering these planets was a major event in astronomy and the world in general, it opened up a greater curiosity. Were there Earth-like planets out there at the proper distances from their stars that could possibly support life? We needed to know!

The "wobble" method utilizing the Doppler Effect led us to discover hundreds of planets and it is still being used today. However, it is not very effective at detecting lower mass planets, like the Earth, which cause miniscule wobbles of their parent stars. There had to be a different technique to uncover the existence of Earth-like planets. Again, we figured it out. The method that can be used to discover any sized planet is called the "transit" method and here is how it works.

First of all, a "transit" is defined as when a smaller body moves in front of a larger body. It is theorized that 1 out of 50 star systems with planets will have their orbital planes tilted to be along our line of sight, allowing us to observe transits of planets in front of their stars. If we had a good enough telescope with a sensitive light meter, preferably in space, we would detect a star's light dimming when one of its planets moved in front of it. The light output may only decrease by a fraction of a percent but it would be measureable. Not only might we be able to know if a planet is in the system, we also could determine if there were multiple planets (from multiple dips in the star's light output), and the diameters, periods (time to orbit), and atmospheric composition of those planets. The diameters, periods, and atmospheric composition are especially intriguing, as we would be able to discover if Earth-like planets are in the "habitable zone" (also called "Goldilocks zone") of their stars. The habitable zone is the region around a star where the temperatures are just right for liquid water to exist on the surface of the planet. Of course, being in the habitable zone does not guarantee liquid water, as other factors, such as atmospheric pressure and mass play a key role. So, now, if we find Earth-like planets in the habitable zones of Sun-like stars, we are on the right track for finding out how many are out there and the chances for extraterrestrial life.

In 2009, we sent up a spacecraft called "Kepler" (named after Johannes Kepler, who in the early 1600s discovered the relationship between the period of a planet and its distance from its star) that had its mission to observe thousands of stars and determine if any had planetary transits. Over the past 7 years, the results have been astounding. It has confirmed over 2,000 planets and has found many Earth-sized ones in the habitable zones of their stars! The Kepler Planet Finder mission has even discovered planets with water vapor in their atmospheres and ones orbiting binary stars (reminiscent to Luke's home world, "Tatooine," in the *Star Wars* science fiction saga). We have also just discovered a planet orbiting Proxima Centauri, the closest star to the Solar System at just over 4 light years away.

Extrapolating the Kepler mission data, astronomers have calculated that there are over 100 billion planets in the Milky Way Galaxy and over 50 billion planets in the habitable zones of Sun-like stars! With an estimated 9 billion of those planets being Earth-like and in the habitable zones of Sun-like stars, how many have life? How many have intelligent life? It is only a matter of time before we find out. And that time may be very short.

*In 1992, there were two exoplanets discovered that were orbiting a pulsar. These exoplanets were actually the first to be discovered but this system was very peculiar as a pulsar is the core of a dead star. These planets probably formed out of debris from the supernova that led to the birth of the pulsar.

Chapter 62

The Dog Star: Beacon of the Night Sky

If you go outside in the winter, between 6:00 and 9:00 p.m., the brightest group of stars you will see are the stars in Orion. Orion is most easily identified by three bright stars in a row that represent the belt of the great hunter. If you draw an imaginary line through the belt stars from right to left, and continue this line leftward and downward, you will encounter Sirius, the brightest star in the night sky. Only the Sun, Moon, and sometimes the planets Venus, Mars, and Jupiter appear brighter than Sirius.

There are two reasons for the great brightness of this star. One is that the star is relatively hot (17,500° F compared to 11,000° F for the Sun). A second reason is that Sirius is one of the nearest stars to the Solar System, a scant 9 light years distant. Only 6 stars are closer to the Sun than Sirius.

The name Sirius comes from a Greek word that means "sparkling" or "scorching," as relating to how it appears. At times, when viewed low on the horizon, through thick layers of Earth's turbulent and prismatic atmosphere, the star sparkles with the spectral array of a precious gem, appearing first red, then blue, then green. At such times, it conjures images of the Christmas star or, perhaps, an alien spacecraft.

In ancient Egypt, Sirius was more than just another pretty face in the sky. At that time, it was noticed that in late June this star rose in the eastern sky just ahead of the sunrise. This "heliacal" rising, as it is called, happens to occur a few weeks before the annual flooding of the Nile, an event vital for riparian fertility and agriculture in that region. The appearance of Sirius, therefore, served to foretell of the annual inundation and was cause for great preparation as well as celebration. While it is a coincidence that the heliacal rising of Sirius coincides with the flooding of the Nile, the Egyptians saw it as cause and effect. They believed that Sirius was the instrument that brought the life-giving flood. They called the star *Sothis* and gave it the title "Mistress of the Year."

Another name for Sirius is the *Dog Star*. This denotes the fact that it is part of Canis Major, the Great Dog. Canis Major and the adjacent Canis Minor are the mythological hunting dogs of Orion. It was the star Sirius, rather than the Sun, that was held responsible for the scorching heat of July and August. The term "dog days of summer" refers to the midsummer time when the Dog Star is closest to the Sun.

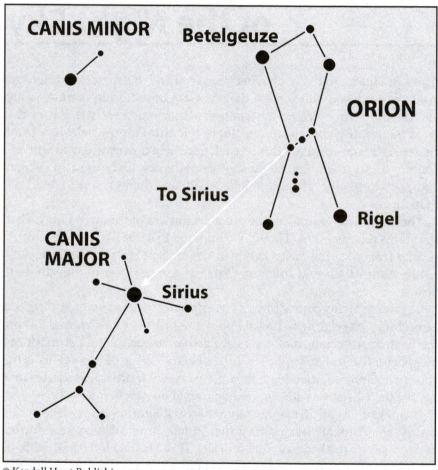

Chapter 63

Hubble Discovers the Universe

At the beginning of the 20th Century, there was an acrimonious controversy in astronomy. The source of this controversy was fuzzy patches of light called spiral nebulae in the night sky. Most astronomers thought they were clouds of stars within our galaxy, local to the Sun. But a few astronomers who had more imagination thought they were massive conjuries of stars seen at a great distance. They called them galaxies or island universes. Which hypothesis was correct?

At the Mount Wilson Observatory, north of Pasadena in California, there was an astronomer by the name of Edwin Hubble. Hubble was interested in these nebulae. He used the great 100 inch diameter reflecting telescope to study them and in particular, one nebula named Andromeda. This is the brightest nebula in the sky. When he focused on the spiral arms of this nebula, he found the kind of star he was familiar with from work he did on The Milky Way. These were Cepheid variable stars. Now these stars can have their distances measured because they pulsate within a given period related to their brightness. In other words, if you see a Cepheid with a certain period of pulsation, you can determine its brightness and hence, its distance. Hubble saw some stars in Andromeda that had the characteristics of Cepheids. These stars seemed very dim when he measured their brightness on the photographic plates he took. But from his experience with other Cepheids with the same periods, they should have been quite bright. The difference between their apparent and absolute magnitude is related to the distance. The distance to the Andromeda Nebula was discovered to be about two and one-half million light years, so the Andromeda Nebula was actually the Andromeda Galaxy! Hubble opened the universe as it went from being a single galaxy (ours) to an entity containing billions of galaxies!

Chapter 64

The Big Bang vs. the Steady State

Georges Lemaitre, a Belgian priest, hypothesized the Big Bang Universe in 1931. In this theory the universe began with a big explosion that started everything expanding, but there were some people who objected to a universe with a beginning. Among these was Fred Hoyle, a famous British astrophysicist. He thought that it didn't make any sense that there was a time when the universe did not exist. To play devil's advocate, he hypothesized the universe always existed and had always had the same properties in space and time. This was dubbed the Steady State Theory. It was based on what Hoyle called the Perfect Cosmological Principle. The idea that the universe was the same at all times and in all places. For the universe to be the same at all times it would have to exist forever.

So coming into the 1960s we have two theories of the universe: The Big Bang Theory and the Steady State Theory. How to choose between the two? The difference between these two theories is that one universe, the Big Bang universe, evolves, and one, the Steady State, does not. The resolution to the controversy came in 1965. It was in that year that a team of astrophysicists from Princeton University decided that they had the technology to search for the Big Bang. How would you do this? Well, if there were a Big Bang, it would have occurred about 13.7 billion years ago. Since in astronomy, looking into the distance is equivalent to looking into the past. This is because of the large distances between objects combined with the finite speed of light. So all we had to do was to look 13.7 billion light years away and the Big Bang would be there. All around us at a distance of 13.7 billion light years there should be radiation coming to us from the Big Bang.

What should it look like? Well, for one thing, this is the remnant radiation or heat left over from the Big Bang so it should be very hot, something on the order of 3000 Kelvin. For another, it should be diluted, due to the distance. Of course, because of the expansion of the universe, any light coming from so great a distance should be red shifted, in this case to a wavelength of about 1 millimeter—a microwave signal. And third, the radiation should have the profile or form of a perfect radiator, a so-called "black body."

Scientists at Princeton started building an antenna to receive this radiation. Meanwhile, unknown to them, two physicists from Bell Labs, about 40 miles down the road, were experimenting with their own radiotelescope. These physicists, Arno Penzias and Robert Wilson, were not astronomers; rather they were experimenting with microwave communications techniques. Their first step was to take their telescope and get a good clear signal from it. But when they used the telescope they found that it had a hum of background interference. Try as they might, they couldn't get rid of the interference. Finally, they were driven to taking the telescope apart and re-soldering all the joints in the antennae. That didn't help either. They evicted the two pigeons that had set up housekeeping in their antenna and went so far as to scrub off the "white, sticky substance" the pigeons had deposited, all to no avail. They mentioned their problem to a friend in the business, who happened to know of the work going on at Princeton. They put the two parties into communication with one another and the mystery was solved.

Penzias and Wilson had discovered the background radiation of the universe, the distant echo of the Big Bang. It is today called the 3-Kelvin Microwave Background because it is consistent with a universe that is radiating at a temperature of 3 Kelvin, just above "Absolute Zero". With this discovery, we have settled the question of which theory is right, the Big Bang or Steady State. The Big Bang demands a microwave background and the Steady State does not, so most astronomers today believe in the Big Bang Theory.

Oh yes! Several weeks later, the experimenters at Princeton completed the construction of their equipment and they confirmed the discovery. Penzias and Wilson, who did not know what they were looking for and did not know what it was when they found it, won a Nobel Prize!

Chapter 65

Gravity Waves

In Einstein's Theory of General Relativity, there is a prediction of "gravitational radiation." Just what is this? In short, gravitational radiation is the energy associated with gravitational waves or the ripples in spacetime that occur when masses accelerate. In this respect, they are the same as electromagnetic radiation that is emitted when charged particles are accelerated. Like electromagnetic radiation, gravitational radiation travels at the speed of light.

In theory, gravity waves are generated when a massive star collapses to become a neutron star or a black hole, or possibly when neutron stars or black holes collide.

The first efforts to detect gravity waves took place in the 1960s when Joseph Weber of the University of Maryland built a gravity wave antenna. The antenna consisted of a cylinder of aluminum about 10 feet long to which piezoelectric crystals were attached. When gravity waves entered the chamber, they caused the cylinder to oscillate and the crystals to give off electric currents. The device was connected to a similar device in Chicago. For a gravity wave to be real, both antennae would have to go off at the same time, proving that the disturbance was not local (e.g., due to a passing truck).

Finally, on September 14, 2015, the detection of gravity waves was confirmed! The discovery took place at the twin Laser Interferometer Gravitational-wave Observatory (LIGO) detectors, located in Livingston, Louisiana, and Hanford, Washington. The gravitational waves detected were thought to be from the merger of two black holes 1.3 billion light years away. At each detector, a 2.5 mile long "L" shaped LIGO interferometer uses laser light split into two beams that travel back and forth down the arms (4 foot diameter tubes). The beams are used to monitor the distance between mirrors which are exactly positioned at the ends of the arms. According to Einstein's General Theory of Relativity, the distance between the mirrors will change by an tiny amount when a gravitational wave passes through the detector. A variation in the lengths of the arms less than 1/10,000 the diameter of a proton can be detected.

This discovery was a monumental achievement and a further confirmation of general relativity.

Chapter 66

It's a Matter of Time

In order to keep time, one needs a periodic motion, one that is, hopefully, very accurate. Then, one needs to count the number of cycles elapsed as a measure of the elapsed time. Historically, the Earth, Moon and Sun have been the most accurate in repetitive motion, so astronomers have become the "keepers of time."

But first, lets define a few concepts related to time. The local meridian is a great circle which goes through the north point on the observer's horizon, the North Celestial Pole, the local zenith and the south point on the horizon. The local meridian is the circle to which we are referring when we say a.m. and p.m. A.m. simply means "ante meridiem" or before the meridian, and p.m. means post or past the meridian. These terms refer to the position of the Sun in the local sky.

The sigma point is the juncture of the local meridian and the celestial equator. It is at a fixed point in the sky and serves as the base point for telling time. As time goes by, objects sweep past the local meridian, driven by the turning of the Earth. The local hour angle, defined as the time elapsed since the object was last on the meridian, is simply the angle between the object and the sigma point. Now at last we're ready to tell time.

There are at least three kinds of time out there, depending on what you use for a reference. In sidereal time, the stars are used as reference points. It could be any star, but we have chosen to use the Vernal Equinox, which lies at the juncture of the celestial equator and the Ecliptic, or path of the Sun through our sky. So, simply stated, the local sidereal time is defined as the hour angle of the Vernal Equinox, and a sidereal day is the period of revolution of the Earth with respect to the Vernal Equinox. Another type of time is solar time. As you might have guessed, this time the reference object is the Sun. Solar time is defined as the hour angle of the Sun plus twelve hours. We add the 12 hours to force the day to start at midnight and not at noon as would otherwise be the case if the definition were simply solar time equals the hour angle of the Vernal Equinox. The only thing wrong with this is that the hour angle of the Sun does not progress uniformly throughout the year. Sometimes it goes faster than at other times.*

To correct for this, astronomers have defined a mean Sun that goes around the celestial equator at a uniform speed. The equation of time relates to the position of the mean Sun with respect to the apparent, or real, Sun. So we have local mean time as being the hour angle of the mean Sun plus twelve hours.

The last correction we have to make is one to account for the fact that every different longitude on Earth has a different mean solar time. We solve this problem by creating zone times. Zone time is characterized by time zones which are 15° wide and centered on time zones at 0° (the Prime Meridian) and, moving westward, at 15°W, 30°W, 45°W, 60°W, 75°W . . . 180° (the International Date Line), and, moving eastward, at 15°E, 30°E, 45°E, 60°E, 75°E . . . 180° (the same International Date Line).

By convention, all people within a given time zone agree to keep the same time; the time of the zone center. For example, we in Sonoma County, California (longitude = 123°W) agree to keep the time of the center of our time zone (longitude = 120°W). So the mean time we keep is 3° ahead of the mean time for our longitude. Since it takes 4 minutes to turn 1°, it must take 12 minutes to turn 3°. So our clocks are wrong by twelve minutes, plus the reading for the date from the equation of time (which could be as much as 16 minutes). Now we see what is the matter with time.

---

*The speeding up and slowing down of the Sun is due to the fact that the Earth is in an elliptical orbit around the Sun. So when the Earth speeds up, the Sun appears to.

Chapter 67

Sagittarius and the Galactic Center

Our Sun is part of a system of 200 billion stars called The Milky Way Galaxy. The galaxy is 100 thousand light years (one light year equals six trillion miles) in diameter and 1 thousand light years thick and has a spiral shape. The Sun is not at the center of the galaxy, but rather 26 thousand light years distant from the center.

Where is the center of the galaxy in our sky? It is along the band of light known as the Milky Way (our galaxy as seen from inside of it) and in the direction of the constellation Sagittarius. So if you stand outside and look toward Sagittarius, 26 thousand light years away, you will find the center. What is at the center? That is a difficult question because there is much gas and dust between us and the center. The Great Rift (a series of overlapping dust clouds) in the Milky Way is evidence of that.

We see to the center by using radio astronomy because radio waves can penetrate the dust. Some astronomers think that there are energetic processes going on at the center, possibly even a super massive black hole gobbling up material. This theory is supported by observations that reveal super massive black holes at the centers of other galaxies, for example, the Andromeda Galaxy. We also have detected stars whipping around a relatively compact, unseen mass of approximately 4 million solar masses at the center of the Milky Way. This is strong evidence as to the existence of the black hole. It is believed that during the formation of galaxies, super massive black holes naturally form as material condenses from the cloud of dust and gas from which the galaxy forms.

Sagittarius is a constellation of the archer. He is also a centaur (a human head and torso attached to a horse's body and legs) and centaurs were good archers. He is one of two centaurs in the sky, the other being Centaurus in the far southern sky. At the feet of Sagittarius lies Corona Australis, the southern crown.

Although Sagittarius is supposed to represent an archer, the constellation looks more like a teapot, the Milky Way rising as steam from its spout.

Chapter 68

Stellar Genesis: The Birth of Stars

Astronomers like nothing more than to think about stars. They think about stars' lives. They think about the birth of stars. They think about the death of stars. This may be considered a difficult topic, because it's very hard to see a star evolve or die. The lifetimes of stars are just too long for us to see any change, and what is there to see, anyway? You can't see into the star to know if there are any changes in conditions there. The answer to this perplexing problem is to use computers programmed with physical theory to predict what would happen to a mass of gas if left to its own devices. This is called "computer modeling."

After an astronomer runs the program, he/she has certain long-term models of the star. The astronomer can tell which one makes sense by comparing it to a color-magnitude diagram of real stars. The color-magnitude diagram represents what the real stars are doing, while the model represents the way the stars do things. Once the astronomer gets a good agreement, that is a sign that the correct laws of physics were picked for predicting how stars evolve.

To get stellar birth, all we need is a cloud of mostly hydrogen and helium gas. Then when the cloud collapses and gets denser, it will become a star. While it's one thing to say this, it's another thing to do it. Just why does a cloud start collapsing, after having been stable since the universe was born? While we're not sure, we have certain ideas. One way that a star can start forming is if a supernova blows up in its vicinity. The radiation pressure from the supernova could pass through the cloud and initiate a compression that initiates star formation. Or, passage of the cloud through one of the spiral arms of the galaxy could initiate a compression of the cloud, and start star formation. Either way, it happens. The cloud starts collapsing, and the star is on its way. When the cloud collapses, molecules fall toward the center of the cloud. The in-falling molecules pick up speed and that causes the temperature to rise. The rising temperature causes radiation to generate. In this way, the so called "proto-star" gets more luminous as the temperature rises. After a time, the interior temperature reaches about 13 million Kelvin, at which point a nuclear reaction called hydrogen fusion initiates. The gamma rays (radiation pressure) from hydrogen fusion, along with the gas pressure from the compressed star, are able to balance gravity pushing inward. Voila! The star is born.

The more massive the cloud, the higher the gravity, and the faster it evolves. This process goes along without outside intervention of any kind.

Chapter 69

The Expansion of the Universe

In 1929, five years after he discovered the existence of the galaxies, Edwin Hubble made another great discovery. He had been, in the years after 1924, doing work on the radial velocities of external galaxies. Combining his work with that of V.M. Slipher of Lowell Observatory, he found that almost all galaxies exhibited redshifts in their spectra. And what's more, if we assume the redshift means radial velocity, almost all galaxies are moving away from us. And furthermore, the farther a galaxy is from us the greater its red-shift or velocity of recession. So we see that the whole universe is expanding, much as a balloon expands as we blow it up.

In 1931, the Belgian priest, Georges Lemaitre, used Hubble's work to formulate a theory about the expansion of the universe. He said that the only way the universe could be in a state of expansion is if there had been an impetus at the beginning to blow it outward. Thus, the Big Bang Theory was born.

There is an interesting story about one of the men who worked with Edwin Hubble on this project. His name was Milton Humason, a mule team driver who hauled the equipment up the mountain when building the 100-inch telescope on Mt. Wilson. He hung around with these astronomers when they were setting up their equipment and when they were doing their work. He exhibited a prescient ability as an observer and had fine mathematical skills. It turns out that he stayed on as an astronomer and was instrumental in the work of gathering the radial velocities of the galaxies that Hubble needed for his project.

Chapter 70

The Milky Way: The Via Lactea

The summer sky is marked by the appearance of the Milky Way, a band of cloudy light encircling the Earth. But what is this band of light? This question haunted the ancients. And what do we do when we have a burning question? We make up an answer. Their answer was that the Milky Way was a river of milk from the breasts of Hera, their head goddess and wife of Zeus. This, however, is not our answer. It turns out after many years of concentrated study that the Milky Way is the edge-on view of our galaxy. In 1924 the true nature of the galaxies was discovered by Edwin Hubble. He found that they were giant conglomerations of stars seen at great distances. We, ourselves, are in one of these galaxies.

Picture a very large wagon wheel, say 100 feet across. If you were sitting on one of the spokes of that wheel, the rim would look more or less like a flat horizontal plane encircling you. This describes the Milky Way. We live in a flattened wagon wheel shaped distribution of stars. We are two-thirds of the way out from the center. The number of stars is in the hundreds of billions. The diameter of this wheel is 100,000 light years and its thickness is about 1,000 light years. The galaxy is typical of other galaxies in space. Within range of our telescopes are more than 100 billion other galaxies, each containing billions of stars. And these galaxies extend out to more than 10 billion light years in distance.

In the universe there are different types of galaxies; they differ with respect to shape. The Milky Way is a type called a spiral galaxy because it appears as a spiral when you look at it face-on. The spiral arms are composed of young, bright stars. Also in spiral galaxies there are vast clouds of dark, obscuring cosmic dust. One of these clouds of dust can be seen in the summer sky as the Great Rift, which lies along the Milky Way in the constellation of Cygnus. In actuality, it's the dust that obscures the farther stars.

Another dust cloud is the Coal Sack in the southern hemisphere.

Chapter 71

The Mystery of the Accelerating Universe: Dark Energy

Astronomers are puzzled. There is a mysterious force out there that makes up almost 70% of the universe. This force makes no sense so far, as it appears to be going against the laws of nature. What is this force and what is it doing? Well, it is known as "dark energy" (astronomers attach the word "dark" to things when they are puzzled), and it seems to be accelerating the expansion of the universe.

Dark energy was first detected in 1998, when observations of super-novae in distant galaxies by the Hubble Space Telescope showed that the universe is expanding faster today than in the past. This was not expected and could not be explained. There are, however, some hypotheses as to what this dark energy is. One idea is that "space" possesses its own energy. So as space expands, this energy does not get diminished and with more space comes more energy. This would possibly lead to an acceleration. Another explanation comes from the quantum theory of matter and states that space contains virtual particles that come in and out of existence. However, this hypothesis would lead to space having much more energy than what has been observed, so it doesn't seem like a plausible answer. It has also been proposed that dark energy is a new kind of dynamical energy field or fluid, or something that is counter-acting to normal matter and energy. This idea, however, doesn't get us any closer to solving the mystery as no experiment or observation supports this so far.

There have been other suggestions that just don't seem to explain what dark energy really is and what causes it. However, there seems to be a connection between dark energy and what happened at the very beginning of the Universe. There is a theory, proposed by Alan Guth, that the universe expanded faster than the speed of light when it was less than a second old. In fact, it is proposed that the universe increased its size by a factor of 10^{26} in a trillionth of a second. This theory is called "Inflation" and explains much about the universe today. However, the interesting aspect here is the "rapid" expansion, which is very similar to what we are seeing with dark energy. Could the same catalyst that inflated the early universe be related to this mysterious dark energy? The answer is yet to come.

Chapter 72

Seeing Into the Past

I feel that astronomers have a certain right to gloat over other scientists in other fields of endeavor. After all, astronomers are the only ones who can directly view the past. How is this possible? The universe is so vast that light travels slowly in comparison to the distances involved. So when we look at a star that's a light year away (in actuality, there is no star this close, aside from the Sun), for example, the light takes a year to get to us (the definition of a light year is the distance that light will travel in one year, or about 6 trillion miles). Therefore, we are seeing light that was born a year ago, or alternately, we are looking one year into the past. This can be put to great use. For example, if you want to study the past history of the universe, all you have to do is look into the distance, and you see the past. Astronomers have done this with the Big Bang. They wanted to see evidence that the Big Bang actually happened. Theory postulated that there was a big explosion that occurred about 13.7 billion years ago. So astronomers looked 13.7 billion light years away, and they found it. It was a dim echo from the past. The cosmic glow of the Big Bang is still alive and well, running around the universe. They call it the 3K microwave background radiation and it is the strongest evidence that the Big Bang actually occurred.

We also use the finite speed of light to prove that the universe is an evolving one. We see more quasars (galaxies with very energetic nuclei) in the distance than we see nearby, proving that there were more quasars in the past than in the present, thereby showing that we live in a changing universe. So much for the Steady State theory, which predicts an unchanging universe.

Chapter 73

The Twin Paradox

The Twin Paradox is a paradox in Special Relativity that goes like this: if one twin stays at home, and the other twin goes on a space trip of forty light years, at .999c (where "c" equals the speed of light), the twin that goes on the space trip will age about five years during the time of the trip. The twin that stays at home will age forty years. Now the paradox: Special Relativity tells us that we can assume any observer is stationary. So why can't we assume that the observer, whom we said went on the space trip, was stationary, and the observer, whom we said was stationary, was actually the one that went on the trip? Then the observer, whom we said aged five years, actually aged forty—and the one we said aged forty years actually aged five. Which is it? This seems paradoxical. Actually, if you're confused, what you are confused about is Relativity Theory. The answer is simple. One twin had to go away from the Earth, stop, turn around and come back. That's the only way we can compare the ages side by side. So the twin who went on the trip had to accelerate. Upon stopping and coming back, he broke the rules of Special Relativity—namely, thou shalt not accelerate. He's broken the rules.

The result here is that the twin who goes on the trip ages less than the stay-at-home twin.

Chapter 74

The Missing Mass in the Universe

There is a problem with the universe. Some of its mass is missing. Ninety percent, to be exact. "What!" you say. "Ninety percent of the mass is missing?" How do we know such a thing? And if its missing, how do we know it was there in the first place? The missing mass question is one of the fundamental problems in astronomy. It all centers around the question of forces, the force of gravity, to be exact. What is the force responsible for holding matter together? Well, that depends on the density of matter. For example, small bodies like atoms and people or rocks are held together by electromagnetic forces. Larger bodies like galaxies and stars are held together by gravitational forces. If the individual atoms in the star, or stars in the galaxy, are moving fast enough, they will escape from the star or the galaxy.

In our universe the matter is lumpy—because the gravity that comes from mass creates a force of attraction. This tends to make matter clump. So stars are created from the gravity of a cloud pulling together the constituent atoms. A galaxy is made from gravity too—the gravity pulling together all of the stars that make up the galaxy. Beyond the galaxy, we have clusters of galaxies, or a "first order clustering," as it is sometimes called. Here, too, gravity causes clustering, but there is a problem.

In the 1920s, Fritz Zwicky noticed that clusters of galaxies seemed to be in rapid motion. The individual galaxies in the clusters were moving too fast. At that rate they would escape from the cluster. He knew the mass of a cluster because he could count the galaxies in the cluster, and he could measure the mass of each one. When he did this, he could compare the speed of the galaxy to the mass of the cluster, and he determined that at the speed of the galaxy, the cluster should have been disbanding. What is the force responsible for holding matter together?

Well, that depends on the density of matter. For example, small bodies like atoms and people or rocks are held together by electromagnetic forces. Larger bodies like galaxies and stars are held together by gravity. So the bottom line was, clusters that should have been disassociating were not. And the problem was the mass of the cluster. He computed that there must be ten times more mass than he saw for the cluster to remain stable. Hence, the missing mass.

The same problem popped up again in a different context in the 1970s. Vera Rubin was measuring the speed of globular clusters. She found that the globular clusters she was working on were traveling faster than she expected. How did she know what to expect? Well, she knew the globular clusters' distance from the center of the galaxy, and she assumed Keplerian rotation for the galaxy. The Keplerian rotation is what we have in the Solar System. The outer planets move more slowly than the inner ones. The other type of rotation is "solid wheel" rotation, where the outer planets move faster than the inner ones (we are using the term Keplerian "rotation" but when addressing the movement of bodies, such as the planets, around a star, the actual term should be "revolution"). It is natural to assume that Keplerian rotation exists when you have things held together by gravity. When you have things held together by electromagnetic forces, you assume solid wheel rotation. The only way Rubin could explain the rapid revolution of the globular clusters was to assume that the mass of the galaxy, heretofore thought to be concentrated within the orbit of the Sun, was now extending about five to ten times the distance of the Earth. This made the mass of the galaxy about seven times, or 85 percent more than they thought it was. Thus, dark matter.

What is this stuff? Astronomers have come to the conclusion that dark matter comes in one of two forms: Cold dark matter, "CDM," or hot dark matter, "HDM." Cold dark matter may exist as one of several items. It could consist of planet sized bodies (MACHOS, massive compact halo objects). Hot dark matter might also be clouds of neutrinos, the most elusive particles known to exist.

Chapter 75

The Shape of the Universe

To ask what is the shape of space wouldn't do much good because you can't picture space with a shape. What if I said that our universe obeys certain rules of geometry. Now we're getting somewhere. You see, there are three types of curvatures corresponding to the three shapes of space, and our universe obeys one of them—and that's the shape of space. All we have to do is find out the rules of geometry that the universe obeys, and we'll know what kind of space we're in.

There are three kinds of geometry. The oldest is "flat" geometry, or Euclidian. This has a curvature of zero. This type of geometry has certain rules that must be obeyed. For example, in flat geometry, the angles of a triangle have a sum of 180 degrees. In flat geometry, the volume of a sphere is 4/3rds πr^3.

But in the 1800s, Riemann found that flat geometry is not the only kind that could exist. He discovered that there was another kind of geometry (now called Riemannian), or positive. In this geometry the rules are slightly different. The sum of the angles of a triangle is more than 180 degrees, and the volume of a sphere drawn in positive space adds up to more than the area that would appear in flat space. As an example of what I'm talking about, consider the surface area of two circles drawn on a flat surface, with the second circle having a radius that is twice that of the first. It is obvious that the surface area of the larger circle is four times that of the smaller circle. Now consider two circles drawn on a positively curved surface (visualize an orange as an example of a positively curved surface). According to the laws of geometry, the area of the entire sphere would be twice that of the area within a circle drawn around the circumference of the sphere (the hemisphere from pole to equator, as it were).

In the 1920s, Nicolai Lobachevski found a third geometry that could be used for this purpose. It's called Lobachevskian geometry, or negative. This type of geometry has triangles whose sum of angles is less than 180 degrees; and a sphere drawn in negatively curved space has a volume equal to more than eight times the volume of flat space.

So our job is to figure out the curvature of space. In general, the astronomer makes an assumption before he does work like this. It's common to assume that objects are distributed uniformly through space, and that the objects mimic the space around us. If you want to observe and measure a certain body of space, you look for enough bodies for a statistical distribution. This means that you can convince yourself that there is a one-to-one correspondence between points on the masses and points in space. Such a survey was done in the last century by radio astronomers from the University of Ohio. It's called The Ohio Survey of Radio Sources. Its results are that the universe was closed. This result is counter to other experiments. It seems now that the universe is open and that it is moving outwards at an accelerated rate.

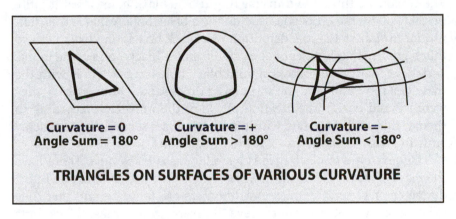

Curvature = 0
Angle Sum = 180°

Curvature = +
Angle Sum > 180°

Curvature = −
Angle Sum < 180°

TRIANGLES ON SURFACES OF VARIOUS CURVATURE

© Kendall Hunt Publishing

Chapter 76

Is the Universe a Black Hole?

The reason we think the universe might be a black hole is that it quite possibly obeys the properties that define a black hole. What is this definition? Simply stated, the definition of a black hole is an object with so much gravity that light cannot escape from it. Black holes come in five varieties. They are as follows: a mini black hole is a black hole with the mass of a mountain and the size of a grain of sand. Quite possibly they were formed during the creation phase of the universe. According to theory, they will evaporate within a fraction of a second due to quickly emitting what is known as "Hawking Radiation."

The next type of black hole is a stellar mass black hole. These have the mass of at least 3 Suns (known as 3 solar masses) compressed into a diameter of 12 to 200 miles. Astronomers are fairly certain that they exist, and can point you to several likely stellar mass black hole candidates.

A third type of black hole is an intermediate mass black hole. These have a mass between one hundred and one million solar masses and have be detected in globular clusters and near the centers of some galaxies.

The fourth type of black hole is called a supermassive black hole, which has a mass of a million to, possibly, fifty billion solar masses, and resides at the center of galaxies. Strange radiations emanating from several galaxies have been attributed to these black holes.

But maybe the most questionable type of black hole is the fifth type—the whole universe. If it is true that light does not escape from the universe, this too would fall under the definition of a black hole. "But we're in it," you say. "Won't we be torn apart by the tidal forces?" It's a feature of black holes that the more massive they are, the less dense they are.

Mini black holes are the most dense. The second most dense are the stellar mass black holes. Supermassive black holes may be less dense than water. They may have a density low enough that you could pass through them without being crushed. They may be considered possible candidates for inter-stellar rapid transit systems. However, at the centers of all black holes lies the singularity, which is a point of infinite, or close to infinite, density. Our known physics breaks down at the singularity so it remains a mystery as to what conditions exist there.

Chapter 77

The Special Theory of Relativity

Newton's "hypothesis" of gravity and "laws of motion" were quickly found by experiment to predict accurately all manner of dynamic phenomena, both terrestrial and celestial. Newton had unified heaven and Earth. The predictive power of his "classical physics" was greater than the world had ever known and he was revered as the genius he was within his own lifetime.

Throughout the next 200 years, physicists and astronomers continued with the laborious task of testing and validating the law of gravity and the laws of motion, as they were then called. The most dramatic confirmation of the laws came in 1846, but the story of the confirmation started in 1781 in the resort town of Bath, England. In that town there lived an amateur astronomer named William Herschel who was short on money but long on ability. To save money, Herschel took to making his own telescopes, which, later studies revealed, were as good or better than any in the world at that time. On the night of March 13, 1781, Herschel, while hunting for comets, discovered a fuzzy disk in the sky where none should have been. Upon further study, it was confirmed that Herschel had discovered a new planet. This was the first time that anyone had discovered a planet and it caused quite a stir. After much squabbling by the English and French, the newly discovered seventh planet was named Uranus.

For the next twenty years, the motion of this planet was observed. Finally, by the early 19th Century, enough data had been gathered to use Newton's laws of motion and law of gravity to compute an accurate orbit for the planet. Armed with this orbit, astronomers predicted the future motion of Uranus. Confident of their mastery of motion through Newton's laws, they sat back to watch the planet confirm their predictions. Alas, it was not to be. First, the planet rode ahead of its predicted motion. Later, it fell behind. But the predicted motion was calculated from Newton's laws. Could it be that the laws were wrong? Not at all!

Between 1843 and 1846, two men, John Couch Adams and Urbain Le Verrier, independently had the same idea. Could it be, they asked, that a heretofore-unknown gravitational influence was affecting the motion of the planet? If there were an as yet undiscovered planet beyond Uranus, its affect was not included in the original computation of the orbit of Uranus. The errors in position, called residuals, would be caused by this undiscovered planet. Adams and Le Verrier used Newton's laws of motion and law of gravity to compute where this planet would have to be to cause the observed residuals. When, in 1846, a search was mounted, the eighth planet was immediately discovered. It was named Neptune and it was found precisely where Adams and Le Verrier had predicted it would be. What a victory for Newtonian mechanics! So complete was the confirmation of the laws of motion and the law of gravity, that most physicists believed that the last word in the physics of motion had been said. Newton was King!

But the flush of victory did not last long. By the late 19th Century, it became apparent that something was wrong with the orbit of Mercury. Newton's laws had predicted that Mercury's orbit would precess. This means that the orbit plane would twist in space causing the perihelion point (the closest approach of the planet to the Sun) to advance around the Sun.

Normally, when one body orbits another with no other gravitational influences around, such a twist would not occur. The elliptical path would repeat over and over again without change. But in the case of the Solar System, there are other gravitational influences around, in particular, the other planets. The effect of the planets on Mercury's orbit is to cause it to precess or twist. These slight additional forces are known as "perturbations," and are well handled by Newtonian physics. In fact, it was the perturbations from Neptune on the orbit of Uranus that led to the discovery of Neptune.

The problem with the precession of Mercury was that, when all of the planets' perturbative effects were taken into account, the magnitude of the precession, predicted by Newton's laws, was 5557 arc seconds/century (1.544 degrees/century). Unfortunately, the observed value of the precession was 5600 arc seconds/century (1.556 degrees/century). Now, this difference of 43 arc seconds/century (0.012 degrees/century) may seem like a small difference to get excited about but the prediction by Newton's law was unambiguous and the astronomical measurements of the actual precession are easy to make and to confirm. There is really something going on here. What could it be?

The first impulse is to guess that we have here a repeat of the Uranus/Neptune situation. Just as Uranus was moving strangely because of the influence of the as yet undiscovered Neptune, maybe Mercury was moving strangely as a result of the effect of an as yet undiscovered planet inferior to (inside of) Mercury's orbit. The planet, prematurely dubbed "Vulcan," was searched for but never found (Vulcan's greatest fame came when it was selected by Gene Roddenberry as the home planet of his character Mr. Spock in "Star Trek"). These 43 arc seconds/century, it turns out, were the harbinger of the downfall of Newtonian mechanics and first rumblings of the relativistic revolution to come.

The goal of theoretical physics is to describe, and therefore be able to predict, the behavior of natural phenomena. Historically, the astronomer has been particularly concerned with the description of the motions of celestial bodies. Other physicists have more generally concentrated on the notions of more familiar bodies such as balls, pendulums, wheels, etc.

These so-called "mechanical events" were the subject of detailed experimentation by Galileo Galilei, while the basic theories of planetary motion are credited to Johannes Kepler.

The "explanation" of natural events, given to us by scientists, may come in various forms. The most efficient language of scientific expression is that of mathematics, but at the time of Galileo and Kepler, sufficient mathematics for the elucidation of their experiments and theories did not exist. It was left to Sir Isaac Newton, who realized that gravity was responsible for the motion of both planets and balls, to invent the necessary mathematics (calculus) and to formalize Kepler's laws of planetary motion and Galileo's mechanical experiments into one simple, elegant theory. When originally published, Newton's theory (today called classical mechanics) was welcomed by the world as the definitive description of the basic laws governing mechanical motion in the universe.

Simply stated, Newton's laws predict the motion of a particle when a force is applied to it. Assume that you are an observer at point O (see figure) watching the motion of a particle of mass M when force F is applied to it. Newton's laws predict exactly how the particle will move.

An important question that scientists must ask is the following. Will the same laws predict the motion of the particle, from the point of view of an observer at O^1, moving with uniform velocity U relative to O?

To answer this question (without actually performing the experiment) one may mathematically "transform" the laws of motion from

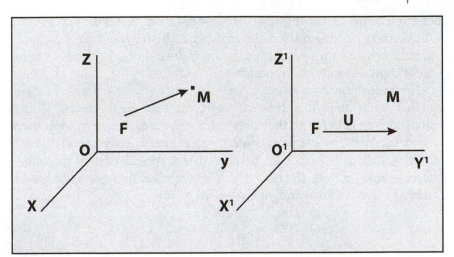

frame of reference O to frame of reference O^1. The mathematical transformation involved is called the Galilean Transformation. When one transforms the laws of Newton with the Galilean Transformation we find that the laws do not change (we say that they are invariant under the transformation). This means that the same laws apply to the prime system as to the unprimed system. We know from experimentation that it is correct that the laws apply uniformly over all frames of reference. But is it the same with electromagnetic phenomenon?

Newton's laws describe mechanical phenomena. The equivalent laws for electromagnetic phenomena are called Maxwell's Equations. What happens when we place Maxwell's Equations in the Galilean Transformation? The answer is they do change form, meaning that you should get a different result from a moving frame. As an example, consider a beam of light. The variant of Maxwell's Equations when you apply the Galilean Transformation means that you should expect a different result in an experiment with light beams when you are stationary as opposed to moving. In particular, you should see a different speed for the beam than a person traveling relative to you would see. But when the experiment was performed by two men named Michelson and Morley, they found the astonishing result that the speed of light didn't change as you change the speed of the observer. Nothing else works like this. The speed of light, by experiment, seems to behave differently than all other phenomena. If you think you understand this, you're wrong, because it truly makes no sense. If a person moving toward you on a spaceship at half the speed of light shines a light on you, you will see the incoming beam as moving at "c" (the speed of light), not 1 1/2 c. If you, in turn, start rushing toward the man with the flashlight at 1/2 c, you will still measure the speed of the beam to be c, not 2c. Speeds of light greater than c seem to be impossible.

Albert Einstein was a young patent clerk in the Swiss patent office. He saw this whole affair in a different way than most professionals. While others tried to get around the observed phenomena, Einstein welcomed it, and made it the prime postulate for a new theory called special relativity. It is called "special" because it is a special case of motion—unaccelerated motion. The new theory was applied to objects moving at constant velocities in a straight line only.

Chapter 78

The General Theory of Relativity

As demonstrated by the Twin Paradox, one has to make sure that one has an unaccelerated system of masses before one applies relativity. In 1905, when the special theory was published, it seemed obvious that unaccelerated motion was relative, but that accelerated motion was absolute. It seemed as though you could tell who was moving in accelerated reference frames. For example, the rotation of the Earth causes a bulge at the equator. The bulge is the evidence for the rotation. It didn't seem possible that you could say that the Earth was still and the universe was spinning around us, as you would have to say if accelerated motion was relative. Another example would be a car accelerating up a freeway on-ramp. The driver is depressed into his seat by acceleration. It seems there is no way to look at this other than that the depression in the seat is evidence for the acceleration.

But Einstein was not happy with this state of affairs. He believed in the symmetry of nature. He thought that the universe would make accelerated motion the same way as unaccelerated motion, that is, relative. He thought about this for ten years, and finally came up with an explanation: the principle of equivalence. This is a postulate to the general theory much as the constant speed of light is a postulate to the special theory.

What is this principle of equivalence? It is the concept that all inertial effects are indistinguishable from gravitational effects. This may seem like a crazy idea, but it is very powerful because it is a postulate that must be taken at face value. Let's see how it works. Before, when we had the problem of the rotating Earth and the equatorial bulge, we said there was only one way to look at it—that the Earth was spinning and the inertial effects caused the bulge. Now there is a new idea. We can also say that the Earth is still, and the universe is spinning around the

157

Earth. The spinning universe would then have to generate excess gravity, which would create the bulge. In the example of the car accelerating up the on-ramp, before, when we had a single way of looking at it we said that the depression in the seat was an inertial effect of the mass of one's body. But now we have an additional way of looking at it. It could be that the excess gravity generated by the universe accelerating backward pulls the body into the seat. So in this situation, again, we have two ways of looking at it because of the principle of equivalence.

As with the special theory, the general theory is more interesting because of the effects it predicts, and in one case, an effect it explains. For years, astronomers have had a problem with the orbit of Mercury. The orbit is not fixed in space, but rather rotates so that the perihelion point advances about the Sun. The perihelion point moves at 5,557 arc seconds per century. Unfortunately, the value predicted was 5,600 arc seconds per century. Now you might think that the agreement was good enough, but the observation, as well as the prediction, is unambiguous. So this difference of 43 arc seconds per century represents a real error that must be explained. Well, Einstein's general theory explains it. It predicts exactly the right number for the precession of Mercury's orbit. It seems that in the vicinity of the Sun, time is slowed down, and this has an effect on the planet that we can observe. The explanation for this observation made Einstein's name a household word in 1916, when the general theory was published.

In 1919, three years after the general theory was published, another experimental verification was made. The theory predicts that a beam of light will be deflected when passing a large gravitational mass. The way to do this experiment is to measure a beam of light when passing by the Sun. This can only be done during a total solar eclipse. So, in 1919, there was going to be an eclipse that fit the bill. Six months prior to the eclipse, the astronomers measured the deflection of a beam of light from a star in a sunless sky. On the day of the eclipse, they measured the deflection with the Sun present. The difference, 1.75 seconds of arc, was the exact deflection of starlight that the theory of relativity predicted.

A third affirmation of the theory of general relativity is the gravitational redshift. This is a statement of the slowing down of time in the vicinity of a large gravitational source. It was confirmed by Robert Dicke and his associates at Princeton University in the 1950s. The idea is that gravity slows down time, so that an object pulsing at a certain rate will slow its rate of pulsation as the gravity increases. This causes a redshift in the spectrum of massive stars.

The general theory of relativity is important because it gives us an alternate way of looking at situations. We can think of any problem as being one of gravity or inertia, and treat it accordingly. This simplifies nature for the scientist.

Chapter 79

Is Time Travel Possible?

Science fiction portrays scenes of time travel. H.G. Wells may have written the definitive book on the subject with *The Time Machine*. Are such trips possible? The concept of time travel presents many logical contradictions. Time travel violates causality, which is the proposition that cause must precede effect. Example: a pitcher throws a ball, and the batter hits it. The cause, which is the pitcher throwing the ball, precedes the effect, the batter hitting the ball. Hitting the ball before it is thrown is in violation of causality. Another example: suppose you get on a spaceship, go back in time, visit your home when you were a boy, then go back further and kill your father when he is a teenager. This violates causality because you kill him before you were born, and that's impossible.

To make time travel possible, there has to be a space-time continuum in which all time exists contemporaneously along with all of space, so that you can go back to a previous time, as well as a previous space. Think about it. You can't have time travel unless all time co-exists. A last thought on time travel. If it were possible, someone would have come back already. If all time co-existed, the future would exist now. Stories like "The Sound of Thunder" are not really science fiction, rather they are fantasies.

Science fiction is supposed to be rational science applied to fiction.

Chapter 80

E=mc²: What Does it Mean, Anyway?

There is a little equation that falls out of the Special Theory of Relativity: $E=mc^2$. What does the equation mean, anyway? Well, a body can have various kinds of energy associated with it. It can have energy of motion, or kinetic energy; energy of position, or potential energy; and it can have rest mass energy. Energy of motion is given by the equation $1/2mv^2$. Energy of position is given by the equation $-gh$, where g is the force of gravitational field strength, and h is the position in the gravitational field. Even if the body is not moving (velocity $v = 0$), and h is zero, you still have a net rest mass energy of mc^2. This is energy that seems to come from nowhere, but the reality is that it comes from mass.

An example will help. The Sun, every second, converts 600 million tons of hydrogen to 596 million tons of helium. Four million tons disappear. In place of it is energy, raw energy in the form of gamma rays. This provides the energy for the Sun. 4×10^{33} ergs (4×10^{26} watt seconds) of energy are produced by nothing more than mass disappearing.

How do we know this is really the way the Sun works? We can't see into the center of the Sun to see the hydrogen disappearing and the helium appearing. What we can do is track a by-product of solar energy generation. Because, in fact, a particle other than gamma rays is produced when the Sun converts hydrogen to helium. That particle is called a neutrino. And a stranger particle does not exist. Before the discovery of neutrinos, the most non-interactive particles were gamma rays. It takes a lead wall three feet thick to stop a gamma ray, while it would take a lead wall fifty light years thick to ensure the stopping of a neutrino (one light year equals six trillion miles). Neutrinos are born in the nuclear reaction that makes energy. Because they are so non-interactive, they pass directly through the Sun, and emerge two seconds after they are born. From there, they make the 93 million mile trip in eight minutes to the Earth, where we seek to capture them. Solar neutrinos, caught from the Sun, are evidence of the reaction that makes helium from hydrogen. The fact that we've only caught about half of the expected number of neutrinos from the Sun does not diminish the evidence that the Sun is generating energy from this process.

Chapter 81

Main Sequence Stars

The "main sequence" star is characterized by hydrogen fusion in the core. Hydrogen fusion is a reaction that demands a temperature of about 13 million Kelvin (23 million degrees Fahrenheit). Usually, on star charts, different colored stars denote different temperatures. Different brightness stars are represented on star charts by different diameter dots. So the main sequence phase is really a phase that demands conditions for hydrogen fusion—that is, a certain temperature. Once a star has gained that temperature in its core, it is bound to be on the main sequence. The temperature comes from gravitational collapse. When the star collapses it heats up. The temperature of the core is hottest because of the great pressure. When the temperature of the core reaches about 13 million Kelvin, a thermonuclear reaction called hydrogen fusion starts.

This is the reaction that stabilizes a main sequence star. Gamma rays born from the reaction rush outward towards the surface. The radiation pressure (with help from gas pressure from the condensed gas) from the gamma rays stabilizes the star against gravity. Each time the gamma rays collide with a proton, they give up a little bit of energy in the exchange. So by the time the gamma ray reaches the surface (approximately 100,000 to 1 million years later), it has lost enough energy to become light and heat. So we see that the gamma rays which stabilized the star are the same rays that make the star shine.

So the main sequence may be defined as the phase of the star when it has reached a certain temperature at the core—the temperature required for hydrogen fusion, or about 13 million Kelvin. Different mass stars arrive at that central temperature with different surface temperatures. This means that the main sequence ends up being a line on the temperature magnitude diagram because the temperature given is surface temperature.

Chapter 82

Nucleosynthesis and Mass Ejection: The Reason We Are Here

Did you ever wonder where chemical elements came from? All 108 elements have a similarity. They're all made up of protons, neutrons and electrons. Protons and neutrons are made of three quarks each, so really, all matter is composed of quarks and electrons. It's like a giant tinker toy. You can make anything out of the parts. All the atoms heavier than helium are made in the stars.

Stars generate helium from hydrogen when they are main sequence stars (see Chapter 81 on "Main Sequence Stars"). A little later, when they are giant stars, they generate carbon from helium. But this material is locked in the core of the star which finally becomes a white dwarf. How do the elements get out of the star? The answer is the S process.

The S process is a way to create elements where they should not be created. Why were no elements ever born in the envelope of a star? The answer is because star envelopes are not hot enough for nucleosynthesis. Well, nature found a way. Nucleosynthesis, the creation of elements heavier than hydrogen, demands that you add protons to an atom to create a high order atom. For example, to create nitrogen, you have to add one proton to a carbon atom, and adding a proton is impossible because of the low temperature of the envelope, and the repulsion of the protons.

The low temperature means that the proton cannot overcome the coulomb barrier and penetrate the nucleus of the carbon atom. But what if we had a neutron? A neutron could penetrate the nucleus of a carbon atom and not be repelled by the coulomb barrier. Then we would have a nucleus with six protons and seven neutrons. Such a nucleus is unstable, and will decay by the process of beta decay.

In beta decay a nuclear neutron splits into a proton and an electron. The electron is ejected from the nucleus, leaving behind the proton, which raises the atomic number by one. The net effect of this whole thing is to take an atom, and by introducing a neutron, raise the atomic number, and therefore the elemental composition of a gas. So nature has done it, performed nucleosynthesis on a gas that has too low a temperature for nucleosynthesis. Now all she has to do is get the gas out of there. This happens routinely in the planetary nebula stage.

The gas in the envelope that contains S process elements (S process is the process of slow neutron capture) is ejected by the giant stars routine mass loss. These gases contain heavy elements, which mix with the pre-existing clouds of gas floating freely in the universe. They enrich clouds with heavy metals. Picture the early Solar System. The Sun is a cloud of hydrogen gas with no heavy elements. If planet formation takes place at that time, there will be no terrestrial planets, because terrestrial planets require heavy materials. But between then and now, stars have formed heavy material from the S process, ejected it through mass loss, and built up the heavy element abundance of the nebula. So when the Sun forms from the cloud, it has a certain mixture of heavy elements. And when the Earth forms from material left over from the Sun, it too has certain heavy materials.

Chapter 83

Is Our Oatmeal Lumpy?: The Structure of the Universe

In 1965, Arno Penzius and Robert Wilson discovered the three Kelvin microwave background—the afterglow of the Big Bang. It was a startling discovery, for it elevated the Big Bang theory above the Steady State theory, the only competition for a theory of the universe. But not too much time went by before the Big Bang theory ran into another challenge. If there was a time in the universe when galaxies formed, there had to be concentrations of matter to act as seeds of the galaxies. These concentrations should show up as slightly hotter places in the 3K backgrounds. Did the universe show evidence for a lumpy 3K degree background?

To answer the question, scientists launched the Cosmic Background Explorer Satellite (COBE) in 1989 to gather high resolution data on the 3K background. As you might expect, the results came in as predicted. The 3K radiation turned out to be quite lumpy, and it gave us evidence from the time prior to galaxies forming. In 2001, NASA launched a more advanced satellite to garner data about the origin of the universe. This satellite, called the Wilkinson Microwave Anisotropy Probe (WMAP) confirmed, with greater resolution, the findings of COBE and gave us data about the age, expansion, and inflation of the universe. We now know we are on the right track with the Big Bang theory and we can account for the formation of galaxies.

Chapter 84

Henrietta Leavitt: Opening the BooK of the Universe

Just because one is an astronomer does not insure against misogyny. At Harvard College in the early part of the 20th Century, women were not allowed to be astronomers. Nonetheless, at Harvard, there were several well-qualified female "computers." Cecilia Payne's thesis on the atmospheres of stars has been dubbed "the most brilliant ever written in astronomy." Annie Jump Cannon is famous for the creation of a spectral sequence, "O, B, A, F, G, K, M." Antonia Maury is famous for her work on a stellar spectra system. Williamina Fleming recorded stellar spectra. Henrietta Leavitt was working on variable stars; this was a class of stars called Cepheid variables. She noticed that certain stars in the Large Magellanic Cloud (LMC) had mean brightnesses that were related to the period (the period being the length of time a star takes to complete a cycle of light output).

Because all the stars in the LMC were the same distance from us, she knew that meant the absolute magnitude was also related to the period. All she had to do was find one star that was a Cepheid variable at a known distance, and she could calibrate the curve. That would give her a period-luminosity relation that could be used to determine the distance to any Cepheid. But there were no Cepheids near enough to have their distances determined in her day, so she used a complex statistical method to do the same thing. Her period-luminosity relation was invaluable in determining the size of our galaxy and the place of the Sun in it. It was also used by Edwin Hubble to determine that other galaxies existed in his mapping of the universe. Imagine what Henrietta Leavitt would have been able to accomplish if she had been allowed to explore as she pleased.

Chapter 85

The Creation of Matter: The First Three Minutes in the Life of the Universe

Based on present conditions and physical theory, astronomers have constructed the standard Big Bang model. This model allows us to compute the conditions of the universe back to 10^{-43} seconds after its creation. Before this time, conditions in the universe were so extreme that current physics is inadequate to give us a clear solution.

From 10^{-43}, to 10^{-35} seconds after creation, we think the universe was dominated by matter-anti-matter annihilation. Antimatter is composed of antiparticles, the "mirror image" of normal particles. For example, an anti-proton has the mass of a proton, but an opposite charge and spin. If a proton and an anti-proton are brought together, they will annihilate one another, converting their mass into pure energy of an amount equal to $E=2mc^2$ (m equals the mass of a proton). There is relatively little anti-matter in the universe today. But from $t = 10^{-43}$ seconds, to $t = 10^{-35}$ seconds, there were 100 thousand antiparticles for every 100,001 particles—a very slight asymmetry. During this period of expansion, matter and antimatter would interact and turn into gamma rays. Similarly, gamma rays would turn into particle antiparticle pairs.

When the universe was about five seconds old, the temperature fell to a few billion K and consisted of protons, neutrons, electrons, neutrinos and photons.

When the universe was about two minutes old, the temperature fell below one billion K. At this temperature, protons fused together to form deuterium ($_1H^2$), tritium ($_1H^3$), light helium ($_2He^3$), and helium ($_2He^4$). The creation of heavy elements is known as nucleosynthesis. These reactions continued for about one minute until the universe was about three minutes old. At this point the temperature dropped below 100 million K and became too cool for fusion. At this point the composition of the universe was 75 percent hydrogen and 25 percent helium. According to theory, any elements heavier than helium were not formed during the Big Bang, but rather by the **triple alpha**, **R** and **S** processes in giant and supergiant stars.

Based on the forgoing, we would expect the first stars to be formed out of a mixture of 75 percent hydrogen and 25 percent helium. This expectation was proved to be correct by studying the population II stars in the halo of our galaxy. These are the oldest stars in the galaxy and their spectra indicate little or no elements heavier than helium.

Chapter 86

Giants and Supergiants

About 5 percent of stars are giants and supergiants. Where do these stars come from? Why, from the main sequence, of course. All stars are first main sequence stars. Giants come from the lower main sequence. Supergiants come from the upper main sequence. The dividing line is about ten solar masses. Stars below ten solar masses evolve into giants, and stars above ten solar masses become supergiants. When a star ends its main sequence lifetime, it grows, powered by a shell of hydrogen fusion. Over the next ten million years or so (faster for more massive stars) the star grows. Finally, at the end of ten million years, the core has shrunken so much that it has become dense enough to be isothermal (the core is so dense that it cannot support temperature differences). The temperature increases enough for helium fusion throughout the entire core at the same time (greater than 100 million degrees Fahrenheit). The star undergoes what is called "helium flash," as the whole core ignites. This is called the "triple alpha process" because a helium nucleus is called an "alpha particle," and there are three of them going into this reaction. This is a giant star—a helium fusion core with a hydrogen fusion shell around it, overlain by a hydrogen atmosphere. The giant star persists in this way for about 100 million years. At the end of this time, the core is completely carbon.

The Sun's outer envelope will be pushed outward by the radiation pressure from the hot shell and the helium fusion core, and will cool to about 3,000 K. It will look red in our sky, which is why it is called a "red giant."

From here the star goes on to the planetary nebula stage, where the radiation pressure from within pushes the envelope outward until it separates from the star and moves into space. The core finds itself exposed to space—this core is called a "white dwarf."

Supergiant stars are another matter. When a main sequence star greater than ten solar masses uses up its hydrogen, it follows the same path as a giant star—up to a point.

When the core is entirely carbon it shrinks to the point where the temperature achieves value high enough for carbon fusion. The giant star did not have enough gravity to create enough pressure for carbon fusion. But the mass of a supergiant being greater than a giant, causes sufficient gravitational pressure to easily overcome internal forces pushing outward. The carbon core shrinks and heats up to carbon fusion temperature and the star ignites causing a slight swelling of the core. The star in this way goes through a series of core fusions. Each fusion creates a higher order element. The star's internal layers of fusion resemble an onion with concentric shells in which higher order elements are being synthesized. In this way the intermediate elements are made.

Radioactive elements are made in a different way. They are created in the supernovae of supergiant stars. The elements are created in the following manner:

Hydrogen and helium are made by the universe during the first three minutes of time. The next elements, and the intermediate elements, are made in giant stars' cores, and in the atmospheres of giant stars by the slow neutron capture process (the "S" process). Finally, the radioactive elements are made by the process of rapid neutron capture (the "R" process) in atmospheres of supernovae and distributed by their violent explosions.

Chapter 87

Planetary Nebulae

The stage of the star after the giant stage and before the white dwarf stage is the planetary nebula stage. This is very short-lived, only about thirty thousand years. Yet we see thousands of planetary nebulae. The driving force behind the planetary nebula is radiation pressure from the fusion shell that exists in the inner boundary of the envelope and the core.

A planetary nebula is a beautiful sight to behold. One of the most famous deep sky objects visible through a small telescope is the Ring Nebula in Lyra (Messier 57). This is a planetary nebula that will probably last about twenty thousand years. Other famous planetary nebulae are the Dumbbell Nebula in Vulpecula (Messier 27), and the Owl Nebula in Ursa Major (Messier 97). The Ring Nebula is the fifty-seventh object in the catalog that Charles Messier compiled in 1774. The irony is that these objects were compiled as objects to avoid while hunting for comets. And yet, they are, as it turns out, about the most interesting objects we can observe. Amateur astronomers spend night after night combing the skies for Messier objects.

Chapter 88

It's a Matter of Spacetime

The number of dimensions needed to describe something is always of great interest to scientists. But first, what is a dimension? Although we use the word routinely in our everyday language, it requires precise defining worthy of a brief discussion.

Simply stated, a dimension is an independent degree of freedom or an independent way to move. Ask yourself the question, "How many independent ways can I move?" Well, there's side to side. That's one. Then there's front to back. That's two. Finally, there's up and down. That's three ways of moving. Any other way would be a combination of two or three ways we've already mentioned.

Another way of looking at it is that the number of dimensions is equal to the number of coordinates needed to specify the location of any place in that space relative to any other. For example, the surface of the Earth is a curved 2-dimensional surface. The proof of this is that it only takes 2 coordinates (latitude and longitude) to specify the position of any place on Earth.

This raises the question: "How many dimensions are there in normal space?" Well, how many degrees of freedom are there? If you say three you'd be right. But wait a minute. Say we're discussing the coordinates of the Death Star. Wouldn't you not only want to know *where* it is but *when* it is? In other words, you'd want to know when the Death Star was at the place indicated by the 3 space coordinates, which, for lack of a better name, we will call x, y, and z. So, to specify a position you also should, for the sake of clarity, specify the time.

The combination of space and time, which is necessary in relativity theory, leads to a phenomenon called *spacetime*. If spacetime is continuous, as is generally believed (black holes are an exception), the universe is a spacetime continuum.

Spacetime was first proposed early in the 20th Century by Hermann Minkowski and made famous by Albert Einstein in his 1905 paper on special relativity. Spacetime intervals, or the lengths of a spacetime trip, while variable in classical mechanics (sometimes called Newtonian mechanics), are invariant in relativity theory. This means that in Newtonian mechanics, from some points of view you get a projection of a space or time trip, so that the length of the trip in space or time depends on the point of view. But in relativity, you get a trip that is the same space time lengths from all points of view. Space may vary, and time may vary, but spacetime is a constant. This makes physics easier to compute. The spacetime interval is called a *worldline*. Worldlines that intersect indicate things that meet.

An *event* in space is a point in spacetime specified by its time and place. Two or more events make a worldline. An example of a worldline would be a straight line parallel to the time axis for a stationary object or a helix for the orbit of the Earth around the Sun.

Chapter 89

Do Neutrinos Have Standing?

There is no more remarkable particle in the universe than the neutrino. Predicted in 1932, the neutrino holds the key to many astronomical conundrums. Search for these particles began when physicists were watching neutrons decay. Neutrons are stable when in the nuclei of atoms, but they decay when outside the nucleus, with a half-life of twelve minutes. Physicists were surprised to find that decay products were emitted in the same direction. This violates rules of conservation of momentum. Usually, when two particles are emitted from a stationary third particle, they will move off in opposite directions—the law of conservation of momentum. So what was happening with these decaying neutrons? Scientists postulated that there was a third particle that was emitted but not seen. An astrophysicist named Enrico Fermi theorized that this particle is neutral and very fast. He called it a neutrino.

Granted, neutrinos have some interesting properties. Because they do not interact electromagnetically, or with the strong force, they can go through matter quite easily. Before neutrinos were discovered, the particle that was most penetrating was the gamma ray, which takes three feet of lead to stop it. By contrast, the neutrino takes fifty light years of lead to guarantee the stopping of it.

Neutrinos are born in the nuclear reaction that makes helium out of hydrogen in the core of the Sun. Three percent of the Sun's energy is released in neutrinos. The neutrinos fly away from the Sun's center near the speed of light, reaching the surface in just two seconds. From there, they fly out into space, reaching the Earth in eight minutes. But they do not stop there. They go through the Earth as though it were not there. They go through your bodies, billions per second.

These days, the most effective ways to detect neutrinos is by using water tanks deep in underground facilities. These tanks are filled with water that contains a large percentage of heavy water (having the hydrogen isotope deuterium). The protons in deuteruim have an associated neutron as part of their nucleus. When a neutrino strikes a quark within a neutron, it causes the neutron to decay into a proton and electron. The electron travels away at a velocity greater than the speed of light in the water, causing a shock wave of faint blue light (known as "Cherenkov radiation"). These neutrino detection facilities are deep underground to filter out other particles (such as cosmic rays) or photons (such as gamma rays). It has to be those elusive neutrinos that are creating this Cherenkov radiation.

Chapter 90

Black Holes, White Holes, and Worm Holes

There is no more exciting object than a black hole. These entities are the stuff of science fiction. In fact, they have been featured in several science fiction stories. What are they? Black holes are objects with so much gravity that light cannot escape from them. Rather, spacetime is pulled into them at a rate faster than light travels. To have so much gravity, you have to have a lot of mass in a very small volume. There are five types of black holes. If you have the mass of a mountain compressed into the size of a grain of sand, you'll get a black hole that will be a mini black hole. Theoretically, when the universe was young, it was a good time for black holes to form.

The most famous type of black hole is a stellar mass black hole. This has the mass of a star (at least three solar masses) compressed into a ball at a minimum of 12 miles in diameter. Another type is called an intermediate mass black hole. These have a mass from about 100 solar masses up to a million solar masses. They may have formed from stellar mass black holes combining. The next type of black hole is a super massive black hole, which may have a million to 20 billion solar masses compressed to a diameter the size of the Solar System. The last type of black hole is the universe. Anything that has so much gravity that the light can't escape is considered a black hole, and that includes the universe.

A white hole is the theoretical other side of a black hole. As black holes collapse inward, white holes expand outward. No one has observed a white hole. It's possible that white holes could disappear into black holes of their own creation. Hence we wouldn't see them.

Until the early 1960s, the case of a rotating black hole could not be worked out. But in 1963, mathematician Roy Kerr worked out the first solution to the rotating black hole. The solution came out to be double valued. What happened is, one plane of existence merged with another, and the interpretation is that one universe joined another. So the interpretation points to the conclusion that a black hole could be a gateway to another universe or a different place in our universe. These theoretical gateways are called "wormholes" or "Einstein-Rosen bridges."

Chapter 91

Supernovae

The most brilliant stars in the sky are supernova stars. These stars are as luminous in one second as the Sun is over its entire lifetime. A single supernova can outshine the galaxy in which it resides. What are these things?

One type of supernova is a result of the catastrophic gravitational collapse of a massive stellar core. The story of these stars is the story of a hunt for the mechanism by which these stars blow up. Astronomers knew that these stars first suffered a gravitational collapse, leading to an envelope that has all manner of nuclear reactions and the high temperature of about 5 billion degrees Fahrenheit. But even with this high temperature the envelope would not blow off. There was something missing. They were stumped. The solution to the problem came with the supernova called SN1987A in the Large Cloud of Magellan. The idea was that when the neutron core was formed, the great pressure squeezed together the electrons and the protons, to form neutrons. This gave rise to lots of neutrinos. The neutrinos took off for distant points. Normally neutrinos can pass through anything. But in this case, the envelope material was so dense that it trapped about 10 percent of the neutrinos. So we have a case where 90 percent of the neutrinos escape from the star, and about 10 percent do not. What happened to the 10 percent that did not? The theory was that they were absorbed by the envelope material, which gave the material the impetus to be blown off. This hypothesis was proved correct in 1987 when neutrinos were discovered coming from the supernova site. They were part of the 90 percent of neutrinos that had escaped. There is also evidence that the creation of black holes at the centers of these dying stars may help cause the supernova.

So now we have a complete theory of supernova explosions which has been borne out by experiment. According to theory, there is a supernova per galaxy every thirty years, but we only see about one out of every ten. The others are being obscured by gas and dust. True to the theory, there have been three supernovae seen in the Milky Way galaxy in the last thousand years. One was on July 4th, 1054. One was in 1572 and the third was in 1604. No supernova has been seen in our galaxy since then (another is overdue). The case of the July 4th, 1054 supernova is of great interest. Ancient Chinese astronomers reported that the star, which was heretofore invisible, became bright enough to shine alongside the Sun in the daytime sky for several months, then faded to non-visibility. These Chinese astronomers recorded where in the sky it was, and when we train our telescopes on that spot today, we find the remains of the star still expanding at seven hundred kilometers per second in the form of the cloud called the Crab Nebula. It is of interest to note that no European astronomer recorded such an event. Perhaps this was due to the belief of Europeans that the sky was changeless.

Supernovae just may be the most important stars in the sky. First of all, they may be responsible for star formation, in that their effect may be to destabilize clouds of gas, and make them form into stars. Also they are responsible for the formation of the heaviest elements, by a process of rapid neutron capture (the R process). Lastly, supernovae are responsible for accelerating hydrogen atoms to relativistic velocities: the cosmic rays which may be responsible for evolution through mutation.

Chapter 92

Harlow Shapley: Galactic Cartographer

One of the great astronomers of the 1900s was Harlow Shapley. He worked at the Mount Wilson Observatory, north of Los Angeles from 1910 to 1920. He was interested in problems relevant to the size of the Milky Way Galaxy, and where the Sun was within it. He was an expert on globular clusters. Globular star clusters are groupings of up to a million stars. There are about 150 to 200 of them in our galaxy. They wheel around the center in long, elliptical paths. They contain some of the oldest stars in the galaxy, being about 13 billion years old. Shapley reasoned that the globular clusters held the key to the size of the galaxy, and the Sun's place within it. He used Henrietta Leavitt's work on variable stars to get the distances to globular clusters. There were about eighty to ninety globular clusters known to him. He made the assumption that the globular clusters formed a spherical distribution around the galaxy that had the same diameter as did the galaxy. This way, by plotting on a graph the positions of the globular clusters, we get an outline of the galaxy. The starting point of the graph is the place where the Sun is. In this way he found that the distribution of the clusters had a diameter of 100,000 light years (his real value was about 50,000 light years because of an error in calibrating the magnitude of the stars). The distance of the center of the Milky Way Galaxy, from the Earth's location within it, was about 30,000 light years.

Chapter 93

A Hundred Billion Stars: Who Counted Them?

It is often said that our galaxy contains a hundred billion stars or two hundred billion stars, or some other number like that. But who counted them? Who has the right to say that the galaxy contains this many or that many stars? Well, when you get right down to it, it's a simple problem. The number of stars in the galaxy is estimated using a very simple procedure. Astronomers determine the mass of the galaxy, and divide that by the mass of an average star. To find the mass of the galaxy, astronomers apply Newton's version of Kepler's 3rd law, which says the sum of the masses in an orbiting system is equal to the cube of the distance between them divided by the square of the period of revolution. They treat the Sun as a point mass orbiting the galaxy, and calculate the mass of the galaxy from there. The Sun is negligible, compared to the mass of the galaxy. Next, they take the distance between the Sun and the center of the galaxy at approximately 12,000 parsecs, and we take the revolution period of the Sun as about 220 million years, and we calculate the answer to be about 10^{11} solar masses. That's the mass of the galaxy.

To get the mass of the average star, we have to survey the stars around us. In the galaxy there are stars of all masses from one tenth of the solar mass, to fifty solar masses. But the mass of the average star is more like a half solar mass. If we divide the last two numbers by one another, we will get the approximate number of stars in the galaxy, and that comes out to be two hundred billion. So when someone asks you the number of stars in an average galaxy, such as the Milky Way, you'll know what to say.

Chapter 94

Star Clusters: Jewels of the Milky Way

In our Milky Way Galaxy, stars occur singly and in groups. The groups come in pairs or sets of many stars. The pairs are called binary stars (sometimes there are triple, quadruple, etc. stars, too), and the groups are called star clusters. Some of them are rich in stars and some are poor. All are beautiful. There are two kinds of star clusters in the galaxy. The first of these is open, or galactic, clusters. These clusters could contain ten to a few thousand stars. There is no real rotational symmetry in an open cluster and these stars are mostly relatively young. The clusters are located in the galactic plane, which is why there are young ones—thousands of them—for they are still being made out of the gases of the galaxy.

The prototype for an open stellar cluster would be the Pleiades in Taurus. This nearby cluster contains several hundred stars and is so young that it still has a layer of dust surrounding the stars (dust shows up as a blue glow around the stars). Its age is believed to be 125 million years. Other open clusters are the Hyades cluster in Taurus and the Praesepe cluster in Cancer.

Also present in the galaxy are globular clusters, which appear in rotational symmetry. They are all old, and inhabit the halo in the galaxy (the place above and below the galactic plane). It is believed that these were the first stars in the galaxy, formed when the galaxy was more of a sphere. Somewhere between 150-200 of these globular clusters exist around our galaxy, and they are rather metal-poor, as you would assume, being old stars—as old as thirteen billion years. The prototype of a globular cluster is the Hercules Cluster, which lies at the naked eye limit in the constellation of Hercules. It contains about a million stars.

Chapter 95

Neutron Stars and Pulsars

Among the glamour stars of astronomy are neutron stars and pulsars. These stars have such bizarre characteristics as to strain credulity.

Neutron stars were first conceived of in the mind of J. Robert Oppenheimer, the head of the Manhattan Project. He wanted to know what would happen if the core of a star, after the star's death, was more than 1.4 solar masses, a condition that was thought to be impossible at that time (it was thought that all cores brought their mass down below 1.4 solar masses to be stable). He found that when the core mass was this massive, the gravity was so strong that the electrons of the core were driven inside the protons, forming neutrons. The result was a star that was a ball of neutrons, a neutron star. Such a thing had never been seen in nature. But did it exist? The astronomical community said no.

There it rested, from the 1930s, when he wrote his paper, until the 1960s, when neutron stars were discovered. The discovery came in a strange way. In the late 1950s, radio astronomy was taking off. Radio astronomers were mapping the skies by using radio wavelengths. It wasn't long before they made a discovery. All around the sky there were intense points of radio emission centered on what optically appeared to be faint blue point-like sources of radiation that were strong in radio wave emissions. These faint blue sources were called quasi-stellar radio sources. But what were they, really?

For the next ten years, astronomers theorized about the nature of these "quasars," as they were called. Finally, the truth emerged. In an attempt to find more quasars, astronomers built a new generation receiver—one with a rapid response time. It just so happened that the same equipment needed to find these quasars would answer the question about the neutron stars. At Cambridge University there was a graduate student named Jocelyn Bell. She was working on quasars using the new equipment, and listening to the incoming signals one night, when she acquired a source of radiation that sounded like rapid radio pulses of about one second in period. When she measured the interval between pulses, she found it was more constant than anything previously seen. Her group puzzled over the signal, and could come up with nothing more explanatory than aliens. So they dubbed the object, "LGM ," which stood for "Little Green Men." But after a few nights, they found another one in another part of the sky. Then another, and another. Whatever these were, they were widespread and obviously of natural origin. They delayed announcement of the discovery for several months, to get a chance to come up with a theory. In the end, the objects were called "pulsars," and the theory emerged thusly:

Picture a core of a star being more than 1.4 solar masses (as you know, 1.4 solar masses is the Chandrasekhar limit—this is the maximum mass for a stable white dwarf). If the core mass is more than 1.4 solar masses, the gravity will overwhelm all other forces and collapse the core. But collapse the core to what? There is only one force left in nature's arsenal—the strong nuclear force. So the star core collapses until the strong nuclear force takes hold, and this stabilizes the star at a diameter of about ten miles. The star at this point is pure neutrons. Conservation of angular momentum makes sure that the star is rapidly rotating (as a skater draws her arms in, she speeds up her spin—this is conservation of angular momentum). It is common to have a magnetic axis tilted to the geographic axis in a planet or a star (in the Earth, the magnetic axis is tilted by 12 degrees to the geographic axis).

When a tipped field is spinning rapidly in a charged field, there is electromagnetic radiation beamed along the magnetic axis. So now we get down to it. The pulsars are rapidly rotating neutron stars. What appear to be pulses are actually continuous beams, crossing our line of sight once every rotation, like a lighthouse beam. The star has to be a neutron star, for anything else would fly apart at their rotation rate. Pulsars are being clocked in milli-second periods. Oh yes! Neutrons have a density of a million billion grams per cubic centimeter (10^{15} g/cm^3). That's pretty dense.

Chapter 96

Life in the Universe, Part I: The Drake Equation

Until quite recently this was not thought of as a serious topic for a serious astronomer. To take part in such a discussion risked being branded as practicing soft science or, at the least, taking part in "fuzzy" logic. All this has changed in the last thirty years. Advances in Astronomy and Astrophysics have taken us a long way toward an understanding of life and life processes, so that having a chapter in a text on astronomy devoted to life in the universe is common.

One of the improvements in this field came from putting the question of life in the universe on a mathematical basis. About fifty years ago, Frank Drake, a U.C. Santa Cruz astronomer, developed an equation to calculate the number of communicative civilizations "out there." Even though it can't give you an exact answer (the independent variables are too poorly known), it does provide a service by succinctly specifying the variables that are important to the problem of determining if there is other life in the universe.

We shall now examine the variables of the Drake Equation, one at a time, and see what value you can reasonably assign to them.

The Drake Equation can be written as:

$$N = n^* \times f_p \times n_e \times f_l \times f_i \times f_c \times L$$

Where:

n^* is the number of Sun-like stars born per year in the galaxy.

f_p is the fraction of those stars with planets.

n_e is the number of planets per star that can potentially support life

f_l is the number of planets with life.

f_i is the fraction of these planets that actually develops intelligent life.

f_c is the fraction of those civilizations that engages in communication.

L is the lifetime of the communicative civilization.

[to be continued . . .]

Chapter 97

Life in the Universe, Part II: The Components

In the preceding document, I specified some of the variables in the Drake Equation but did not give them any values. This I will do now.

Perhaps the best known parameter is n^* (the number of Sun-like stars born per year in the galaxy). A generation of astronomical work has gone into determining the value of this number. For its value, Frank Drake used between five and ten.

The next term, f_p (the fraction of stars that have planets), is better known than the later variables, and is assumed to be about one-half. We have discovered in recent years numerous gas giants using Earth-based techniques and thousands of planets of various sizes from the Kepler Planet Finder Mission. However, most stars in our galaxy are red dwarfs that have little ultraviolet radiation, which provides the spark for life. So n_e (the number of planets per star that can potentially support life) is problematic. Frank Drake, in a landmark paper, chose two planets per star, f_i is a fraction of the above planets that actually go on to develop intelligent life. For the Earth, life developed almost as soon as conditions became favorable, suggesting that biogenesis may be relatively common once the right conditions appear.

But why didn't it happen more than once? All terrestrial life forms stem from a single incident, after all. Frank Drake finally succumbs to this one, and uses the number one for f_l (the fraction of the planets that will develop life), f_i (the fraction of planets that will go on to develop intelligent life) is again marked by the fact that the Earth only developed one intelligent life form and it took 4 billion years, and therefore Drake puts 1 percent down for this term.

f_c (the fraction of planets where the life forms are able to communicate over interstellar distances) is likewise considered to be very rare when you take the Earth into account. Life forms such as whales, dolphins and the like are simply unable to take part in communication with an alien race, even though they may be intelligent. Drake's value for this term is again 1 percent. All that remains to determine is L (the lifetime of the civilization in years). Taking the Earth as an example, there are many reasons to believe that the value of L would be very small. The problems that beset us are extreme, to say the least. What would you choose as the value for L? Drake might have been optimistic when he chose ten thousand years for a value of how long our civilization would last. All told, the values for our equation turn out to be the following: $N = 5 \times .5 \times 2 \times 1 \times 0.01 \times 0.01 \times 10{,}000 = 5$. So, possibly, five planets are out there in our galaxy somewhere with civilizations with communicative capacity. But remember, there is much uncertainty and speculation in this equation. The value could change drastically if scientists change any one of the variables. Most of the variables are close to unknown. Nonetheless, the equation does speak to the terms that are important to consider when confronting this very interesting question.

Chapter 98

Life in the Universe, Part III: SETI

We've convinced ourselves that there are about N extraterrestrial civilizations with whom we could communicate. Where are they? Throughout all of human history, we've not run across even one civilization. This might make you believe that they're really not there. On the other hand, there are all these reports of UFOs. Could this be them? Astronomers take a dim view of UFOlogy. Here's the reason why. An alien civilization is unlikely to resort to going "out there" to reconnoiter other aliens. The distances are so great that it would take a high percentage of one's GDP to send ships to all the planets that you would have to send them to. And consider this—at N civilizations per galaxy they would be x light years apart. How old are we as a communicating civilization? We are about eighty years old. That is how long it's been since the first TV waves went out into space, and since radio and TV are what makes us unique and bright, that's what they would zero in on. In that case, our waves are only about eighty light years out there, well short of the x light years they are from us.

But, how are we to search if not to go there? The answer is a method that is fast, cheap and good. It uses electromagnetic radiation as our emissary. We send radio waves, which reach many stars at once. This was the method that Frank Drake used in Project Ozma in 1960, when he did the first SETI, "Search for Extraterrestrial Intelligence." He was not successful, primarily because he did not survey enough stars.

Every element has a natural frequency of transmission. It is believed that aliens would pick their frequency of transmission carefully. One good choice would be somewhere between the frequency of water and the frequency of hydroxide, the frequency at which the galaxy transmits, the so-called "Water Hole." Just as animals gather around a water hole, alien civilizations might meet around the Water Hole.

And, if you're wondering whose civilization would be most imma-ture, the answer is, it would be ours. Since we have joined the ranks of communicating civilizations just eighty years ago, that makes us just a few tens of years old—compared, possibly, to another civilization's thousands of millions of years.

Maybe we should just keep our mouth shut.

Chapter 99

White Dwarf Stars

White dwarfs are stars that shouldn't exist, all things being equal. They are the cores of giants after the planetary nebula phase. They were first discovered by Friedrich Herschel in 1783. Herschel was observing the main sequence star 40 Eridani when he found a star that was very faint and of spectral class A. Why would a star be so dim, yet so hot? This suggested they were small. Today we know that these stars are about one hundredth the size of the Sun, or about the size of the Earth. They have approximately the mass of the Sun, but their small size makes them extremely dense (10^6 grams per cubic centimeter). The next white dwarf discovered was Sirius B, which orbits Sirius A every half century.

Such a mass has a specific name—"degenerate matter." When mass becomes that dense, it suffers a kind of electron degeneracy, where the electrons are squeezed to the limit, gravity causing the compression. These white dwarfs have strange properties. Their size is governed by their mass, in that the more massive they are, the smaller they are. This is the reverse of normal matter, where things are bigger the more massive they are. A 1.4 solar mass white dwarf is as massive as a white dwarf can be. This is known as the Chandrasekhar limit. Beyond this mass the structure of the star cannot support its weight, and the star collapses. So anything more than 1.4 solar masses and the star becomes a neutron star. A white dwarf can only do one thing. It is so hot that it cools off, slowly giving its energy to space. Because it is very dense, it cools off slowly. A typical white dwarf will take about a hundred billion years to reach the temperature of space. Since the universe is only about 13.7 billion years old, these white dwarfs have not cooled to black dwarfs yet.

Chapter 100

Quasars: Looking Back to the Birth of Galaxies

An interesting case is the case of the quasars. In 1950, after World War II, radio astronomers got back to the business of mapping the universe. Radio astronomy was discovered in 1936, so astronomers had little time to do their work before the onset of the war. The first thing the radio astronomers did was to make a survey of the sky at radio wavelengths. The only curious thing they found were many point sources of radio waves. Astronomers wondered what these were. Being radio astronomers, they had no access to large telescopes, so some from Cal Tech went to the optical astronomers from Cal Tech to ask them to survey these particular positions in space. What the optical astronomers found were dim pinpoints of blue light, one for every position. Now star-like objects giving off radio waves were unknown, so these were very special objects. The spectrum of these showed lines of no known element. Thus, we had quite a mystery on our hands.

When he was driving home from work, Maarten Schmidt, an astronomer from Cal Tech, had an idea. What if the spectrum of these mysterious objects (that came to be known as quasars—short for quasi-stellar radio sources, or QSR) was a normal spectrum, highly redshifted (Redshift indicates recession. It is something that happens in the spectrum of light, which is a pattern of lines. Redshift occurs when the pattern moves as a whole to another place.)? When he got back to school, he measured the displacement of the lines, and sure enough there were lines of normal hydrogen, indicative of ultraviolet. The reason they hadn't been seen before is that no one expected to have ultraviolet lines in the visible range of the spectrum.

Now they had another mystery. If you took the lines to be Doppler shifted, you had the largest Doppler shift ever seen, indicating it was moving away from us at 30,000 miles per second. Certain astronomers objected to this on the grounds that it would take a very powerful source of energy to accelerate all the objects they were seeing to that speed. A dichotomy formed in astronomy about the nature of quasars. On one hand, you had the cosmological people, who felt the quasars were at far distances, and their velocity was cosmological, meaning that they were partaking of the expansion of the universe. On the other hand, some folks thought the quasars were local, possibly ejected from our galaxy. The winners of the debate were the cosmological folks. It turns out that quasars are distant galaxies seen in the process of formation. When a galaxy is born, the massive infall of material forms a massive black hole. When a quasar is young, there is a lot of material for it to eat making it very active and luminous. When it is older, it settles down and doesn't eat so much. Astronomers hypothesize that surrounding an old quasar is a normal galaxy, and this is what they find. When they block out the light from a normal quasar, they find the light from normal stars coming from the region around the quasar.

The mystery of a quasar is solved.

Chapter 101

Triumph in Greece: The Invention of Science

The roots of science are steeped in mystery and shadow. They go back about three thousand years, to the time of ancient Greece. At that time, people saw ghosts and demons in shadowy corners. The first light of science came with the Pythagoreans, circa 600 B.C. These early scientists had a notion that the happenings of nature were explainable by human intellect. What's more, they believed that scientific happenings were governed by rules of nature that even Nature herself had to obey. They thought the world was governed by ratios of whole numbers, such as the four elements (earth, air, fire, water), and discovered the dodecahedron, a twelve-sided figure made up of pentagons, which they attributed to the fifth element, or "quintessence," of which the universe is made. They knew of the sphericity of the Earth, and the fact that it rotated. But there were problems with their philosophy. They could not abide any number of universal constants that were not whole numbers or ratios of whole numbers. This is why they suffered the crisis of philosophy when they discovered irrational numbers—numbers that repeat unendingly, like the square root of 2.

More and more, the Greeks came to rely on observation to verify scientific truth. The one exception to this rule was Plato (circa 400 B.C.), who single-handedly set back Greek science about a thousand years. He ordained that the motions of the heavens were spherical, and that the celestial bodies orbited in perfect circles about the Earth. He was wrong on all counts. The celestial bodies did not revolve in perfect circles and they did not orbit the Earth. The effort to counteract his philosophy took about two thousand years.

Things were a bit better in the Earth sciences where a man named Eratosthenes, who was director of the great library of Alexandria, used sticks to measure the circumference of the Earth to an accuracy of better than 5 percent. He heard a tale told by a traveler that on the longest day of the year, June 21st, the Sun shone directly down a well in the center of the frontier town of Syene. Taken at face value, it is a simple story—but he didn't take it at face value. He thought about this, and asked the question, "Would sunlight do the same thing in Alexandria?" He resolved to try the experiment. When he did, he found the sunlight did not go straight down the well at Alexandria. Rather, it cast a shadow, which turned out to be an angle of seven degrees. So we have the following situation: at the same time, at separate points, Syene and Alexandria, a stick casts no shadow in Syene, but does cast a shadow in Alexandria, amounting to an angle of seven degrees. "This he used to determine the radius of the Earth because if the sticks were vertical in both places that means that the distance between Syene and Alexandria is one fiftieth the circumference of the Earth. He got the number fifty by seeing that the seven degrees is 1/50th of a circle, thus the circumference turns out to be 25,000 miles, very nearly in accordance with reality.

One place where the Greeks fell short was in deducing the nature of our orbit around the Sun. They were trapped by Plato's statement that the Earth was the center of all motion. What they did was busy themselves trying to figure out how the geocentric orbit worked. Callipus, Aristotle, and others spent years working on a false problem. The idea was to find a way to make the orbit work. Finally, a man named Ptolemy came up with the solution—a system of epicycles and deferents allowed for the prediction of planetary motion. The theory of Ptolemy held sway for fifteen hundred years, mostly because of the sponsorship of the Catholic Church. But like any flawed model, this one too broke down. It was off by a slight amount each day—too small to be seen. But after many days had gone by, the error accumulated to become recognizable in the sky.

Chapter 102

"Let's Try the Sun at the Center": The Copernican Revolution

The Ptolemaic geocentric universe held sway for 1500 years until the time of Copernicus. This was a tribute to Ptolemy and his genius in making a model which could predict planetary positions yet be dead wrong. But by the 1100s, the error in the model was apparent as the discrepancy of the predicted planetary positions and their actual locations could be noticed with the naked eye. So astronomers had to construct correction tables to apply to the planetary positions. The situation was inconvenient and cumbersome. In 1543, Nicolaus Copernicus decided to tackle the problem. As a first approximation, he decided to try to fit the Sun in the center of the Solar System. He had good reason to do this. He had done a calculation on the mass of the Sun and determined it to be 6 times more massive than the Earth. Despite the fact that his calculation was wrong (the Sun's mass is approximately 330,000 Earth masses), this did set him in the right direction. He put the Earth in place of the Sun between Venus and Mars. There he stopped. He did not take the necessary step of changing the shapes of the perceived circular planetary orbits. He kept the shapes of the planetary orbits as the same circles that Plato had ordained. He was afraid to change any more for fear of alienating the Church. He didn't even want his theory published, but a friend took it and published it anyway. This published work was titled "De Revolutionibus Orbium Coelestium" (On the Revolutions of the Heavenly Spheres). The friend, in order to save Copernicus any embarrassment, inserted a preface in the book saying that the reader should not take this work as fact. The reader should just think of it as a method of computation. The tale goes that Copernicus received the first copy of his book as he lay on his deathbed.

It should not surprise you that the book was not received very well. After all, it was incorrect. You see, the most serious error in the Ptolemaic hypothesis was the shape of the orbits. They weren't actually circles but ellipses. Copernicus had changed the object at the center of the orbits, which was important philosophically, but not the shape of the orbits which would make the model work. Therefore, no one accepted the model. Not only did it violate church teachings, but it worked no better than the old model. One hundred years would go by before the works of Galileo, Brahe, and Kepler solved the conundrum.

Chapter 103

Kepler, Brahe, and Galileo to the Rescue

There was a crisis in astronomy in 1600. Two world systems were vying to be the one that ruled in explaining the cosmos. The Earth-centered system was championed by the Church and by conservatives. The daring new Sun-centered system was championed by a bunch of upstarts. Who was right? One concerned astronomer named Johannes Kepler wanted to know the mind of God. As a young man, Kepler had experienced a vision of God as a geometer who made the world on geometrical terms. Kepler's life was spent in pursuit of this dream. Also, while he was young, he had a vision of the mathematical model upon which the universe was based. This model was a series of nested solids (six polygons, each inside of the next larger) that represented the orbits of the planets. He spent many years of his life trying to prove this fantasy, yet he could never quite get his model to work. If only he could get the data detailing the planetary motions as they existed in reality, he could try different orbital curves to see which ones would fit. As luck would have it, he had just received a letter from the greatest observational astronomer of the time, Tycho Brahe, Astronomer Royal for the Holy Roman Emperor in Prague. Brahe invited Kepler to join him in his work. Brahe was a colorful person, very different from the serious Kepler. His court had a circus atmosphere and Kepler found himself the butt of all manner of jokes. They argued incessantly and many times Kepler was on the verge of leaving, but then they would reconcile. After about a year of working together, Brahe fell into a stupor and died. Some believe this was due to his overindulgence in food and drink. Others feel that Kepler poisoned him. In any event, with Brahe out of the way, Kepler conspired to get Brahe's data on the position of the planets. This data told him where the planets were at various times in the past. So, he was able to construct a time baseline from the previous

20 years. He experimented with various orbital curves to see which would fit his data. He tried an ellipse (he had tried it once before but made a mistake in the computation which caused him to dismiss it) and it worked perfectly! After this discovery he published his 1st law: "The shapes of the orbits of the planets are ellipses, with the Sun at one focus."

Kepler's 2nd law, also known as "The Law of Areas," pertained to the varying speed of our planets' orbit around the Sun. This law is stated as: "A line from a planet to the Sun sweeps out equal areas in equal times."

Years later, Kepler formulated his 3rd, or "Harmonic Law." This law related a planet's distance from the Sun with the time, or period, it took to orbit the Sun: "The squares of the sidereal periods of the planets are proportional to the cubes of their semi-major axes (average planet-Sun distance)."

Kepler had answered the question, "What are the true shapes of the orbits of the planets?" Tycho Brahe was instrumental, in that Kepler couldn't have done it without his data. In the same span of time, Galileo Galilei was working on the same problem, which he went about solving in a different way. He, too, was interested in figuring out whether Copernicus or Ptolemy was right. He heard about the invention of something called a telescope and quickly built one himself, and with it made monumental discoveries about the heavens. Everything he looked at seemed like a new object. He looked at the Moon and saw deep craters. He looked at the Sun and saw spots. Everything was supposed to be perfectly smooth and unblemished, so these sightings represented failures of the Platonic model. If Plato were wrong about this, why couldn't he be wrong about everything? But the most important observation he made was that Venus went through all phases, varying in apparent size. This observation about Venus indicated that it had to be orbiting the Sun. He also observed that Jupiter had satellites which revolved around the giant planet. These satellites were also violating the Platonic model in that they were going around Jupiter and not the Earth. Galileo's observations got him into major trouble. He was hauled off before the Inquisition and forced, by threat of torture, to recant his discoveries. Even worse, he was forced to spend his remaining years under house arrest.

Chapter 104

God said, "Let there be light," and There was Newton

Isaac Newton was born on Christmas day in 1643. His father had died several months before he was born. His mother's new husband did not get along with young Isaac, so he was brought up by his grandmother. All the while his mother lived in sight of his family farm at Woolsthorpe Manor in Lincolnshire, England. He showed early promise in math and physics which made his uncle intervene for him at Cambridge University. He matriculated there in 1660.

In 1665, he graduated with a degree in math. In the same year, the governors closed the school because the plague swept through London, 70 miles south. The students were sent home and Newton went to the family farm to ride out the storm.

He spent the next 18 months thinking of and doing different experiments. This period of time is known as the "Miracle Year" in physics. During this time he invented whole fields of endeavor, any one of which would have insured his everlasting fame. He developed the Law of Gravity, the Laws of Motion, Calculus, the Newtonian telescope, the basic laws of spectroscopy, and proof of the Binomial Theorem.

In 1667, he went back to school and accepted a professorship in mathematics. He joined the Royal Society where he published a paper on the proof of the Binomial Theorem, which is a method for multiplying out mathematical expressions in the format $(a + b)^n$. The theorem was criticized by some members of the society, which was a very reasonable response. Newton, however, was sensitive and he vowed not to publish any more work. So all of his other great discoveries went unpublished for many years.

One day, twenty years later, his friend Edmond Halley came knocking at his door for help in attacking a problem in physics. Halley wanted to know what would be the orbit of a comet under the single force of the Sun.

"That's easy," said Newton. "It would be an ellipse."

"How do you know?" said Halley.

"Why, I did it twenty years ago," said Newton.

"Can you reproduce the data?" said Halley.

"Certainly," said Newton.

When Halley next went to see Newton, he found the proof of the orbital shape of comets waiting for him. He asked Newton what else he had done twenty years earlier that hadn't been published. When Newton told him, Halley became exasperated, vowing to himself he would force Newton to publish his works. Newton had an enemy named Robert Hooke in the Royal Society. Halley used Robert Hooke as bait to get Newton to publish his book, telling him that Hooke was laying claim to Newton's work.

Halley published Newton's book, *The Principia or The Principles of Mathematical Physics*, at his own expense. It is is considered the finest work in physics and math ever to be published.

In later years, Newton became the Director of the Mint, presiding over the changing currency. Nonetheless, he is one of those rare individuals who was recognized for the genius he was throughout his lifetime.

Chapter 105

Pinwheels of Stardust:
The Structure of Galaxies

The Milky Way is a barred spiral type galaxy in the Hubble classification scheme. That means it has spiral arms. The structure of a galaxy depends very much on the initial conditions that created the galaxy. Ours is no different. When a galaxy starts condensing out of a cloud in inter-galactic space, the first objects that form are globular clusters, having somewhat of a spherical distribution. Because they are the earliest objects, the very oldest objects—their ages range from 6 to 13 billion years—they are called Population II objects, and they have highly elliptical orbits. This spherical volume of space around the galaxy is called the galactic halo or corona. There is a thin distribution of single stars in the galactic corona. They are all metal-poor stars. We find two types of variable stars, W Virginis and RR Lyrae, moving through our galaxy's halo at velocities in excess of 70 miles per second. RR Lyrae stars are giant stars that pulsate in both diameter and brightness in periods of about 0.5 to 1.5 days.

As the galaxy continues to collapse under gravity, its rotation speeds up, due to the conservation of angular momentum. Fast rotation produces turbulence in star forming regions of dust and gas, slowing down the star formation process. This, in turn, means that the gas and dust last longer in the disk of the galaxy. Objects that continue to form in the disk of the galaxy are called Population I objects. They are characterized by open (galactic) clusters which are zero to several billion years old. The last things to form in the galaxy are the spiral arms. These giant pinwheels of gas and dust are outlined by newly formed, ultra hot stars called O-B associations. These hot stars emit ultraviolet light that ionizes nearby gas into fluorescence. Such areas of glowing gas are called HII regions. These are the youngest parts of our galaxy and are found throughout its spiral arm structure. This scenario probably repeated ad infinitum in all of the galaxies in the cosmos.

TIDBITS GLOSSARY

ACCRETION #20: The pulling together of a mass (usually gas) because of self-gravity.

ADAMS, JOHN COUCH #53, 77: In 1846, used Newtonian mathematics to predict the position of a hypothetical trans-Uranian planet from the residuals of Uranus.

AIRY, SIR GEORGE #53: English astronomer and mathematician who established Greenwich as the location for the Prime Meridian; (1826) Director of the Cambridge Observatory.

ALBIREO #7: Double star system of magnitude 3.5 and 7; one of the finest visible pairs, gold and blue in color.

ALDEBARAN #8: One of the brightest stars in the constellation Taurus representing one of the fiery eyes of the bull. Aldebaran means "The Follower" (of the Pleiades).

ALDRIN, BUZZ #30: Apollo XI astronaut. Second man on the lunar surface.

ALGOL #4: Eclipsing binary star, the second brightest in Perseus. Represents the head of Medusa.

ALPHA CENTAURI #43: Closest star system to the Solar System, 4.3 light years away.

ALPHA PARTICLES #58, 86: Stripped helium nuclei.

ALTAIR #7: Star of magnitude 1.3 located at the head of Aquila the Eagle. One of the stars in the Summer Triangle.

ALVAREZ, DR. WALTER #27: Co-discoverer of the iridium layer and the impact theory of the mass extinction that occurred between the Cretaceous and Tertiary periods.

ANDROMEDA, THE STORY OF. #4: The Perseus-Andromeda-Pegasus story in Greek mythology.

ANTARCTIC CIRCLE #23: Small circle at latitude 66.5 degrees South, within which there are 24 hours of daylight on December 21st.

APIS #8: Name of the living bull god of ancient Egypt.

APOLLO #6, 7: In Greek mythology, the son of Zeus and Leto, and brother of Diana.

APOLLO 11 #30: First manned exploration mission to accomplish a lunar landing.

APOLLO 13 #30: Apollo mission which suffered an explosion in service module.

AQUILA #7: The constellation of the Eagle.

ARCTIC CIRCLE #23: A circle of latitude 66.5 degrees North of the Earth's equator.

ARCTURUS #2: The brightest star (apparent magnitude = −0.04) in the constellation of Boötes.

ARIEL #51: A satellite of Uranus.

ARISTOTLE #59: (384 BCE–322 BCE) A Greek philosopher, teacher of Alexander the Great and student of Plato.

ARMSTRONG, NEIL #30: Apollo 11 astronaut and first man on the Moon on July 20, 1969.

ASTERISM #1: A figure in the sky, which is not a constellation, but part of one; or a figure made from parts of several constellations.

ASTEROID BELT #24, 41, 47: The zone of minor planets extending from the orbit of Mars to the orbit of Jupiter, about 50 AUs thick.

AU #32: A unit of measurement; the mean distance from the Earth to the Sun, rounded off as 93,000,000 miles.

AUTUMNAL EQUINOX #18, 34: When the Sun crosses the Celestial Equator traveling from north to south. It usually occurs on September 22nd although it may vary by plus or minus one day.

BAILEY'S BEADS #42: Bright flashes of light seen around the Moon at second contact of some solar eclipses.

BELL, JOCELYN #95: Astronomer who discovered pulsars in 1967.

BELLATRIX #6: A star at the left shoulder of Orion the Hunter.

BETA DECAY #82: A type of nuclear reaction in which an electron is ejected from the nucleus of an atom, raising the atomic number by one.

BETELGEUZE (ALSO BETELGEUSE) #5: A magnitude 0.4 star at the right shoulder of Orion the Hunter.

BIG BANG THEORY #64, 69: The theory that the universe started X number of years ago from a highly compressed state and has been expanding ever since.

BIG DIPPER #1: A northern hemisphere asterism that looks like a ladle.

BLACK DWARF #99: The last stage of stellar evolution when a star totally loses its kinetic energy.

BLACK HOLE #65, 76, 90: A stellar core with so much gravity that light cannot escape its gravitational field. Also, a region with an escape velocity of greater than the speed of light.

BLOOD MOON #16, 31: The first full Moon after the Harvest Moon, also known as the Hunter's Moon.

BODE-TITIUS PROGRESSION #41: The equation (n + 4)/10 = distance from the Sun in AU. A progression of numbers indicating the distance from the Sun.

BOÖTES #2: Northern hemisphere constellation of the "Bear Driver."

BRAHE, TYCHO #102, 103: 16th Century astronomer who supplied Kepler with positional data on the orbit of Mars, enabling Kepler to prove that the planet moved in an elliptical path.

CALCULUS #104: A mathematical theory invented by Isaac Newton, used to divide quantities into a near-infinite number of infinitely small rectangles which may be subsequently re-added.

CANYON DIABLO METEORITE #24: Any iron meteorite which comes from the Canyon Diablo area of central Arizona.

CAPTURE #20: A theory of lunar origin which has the Moon being captured from someplace else in the Solar System.

CARBON DIOXIDE #37: A molecule containing one carbon atom and two oxygen atoms.

CAUSALITY #79: A principle of physics that says that the cause must precede the effect.

CELESTIAL EQUATOR #28, 66: A great circle on the Celestial Sphere which is everywhere 90 degrees from the North Celestial Pole.

CELESTIAL SPHERE #28, 34: A sphere of infinite radius around the Earth.

CENTAUR #67: A fictional creature half man and half horse. Centaurs were supposed to be very good archers.

CEPHEID #63, 84: Pulsating variable stars with periods of 3 to 100 days.

CERES #41: Formerly the largest "minor planet" in the Solar System with a diameter of 590 miles. Was promoted to "dwarf planet" status in 2006.

CHALLIS, J. #53: The director of the Cambridge Observatory in 1846.

CHANDRASEKHAR LIMIT #99: The upper mass limit for a stable white dwarf. M_c=1.4 solar masses.

CHARON #39, 57: One of the five known satellites of Pluto.

CHICXULUB #27: Location on the Yucatan Peninsula where an asteroid hit 65 million years ago.

CHROMOSPHERE #58: A region of the Sun found immediately above the photosphere, about 2,500 km thick.

CLASSICAL PHYSICS #77: The term normally applied to the physics of Isaac Newton.

COAL SACK, THE #70: A dark cloud of obscuring matter in the southern sky.

COLLINS, MIKE #30: Service module pilot on the Apollo XI mission.

COLLISION HYPOTHESIS #20: Theory of the origin of the Moon.

COLOR-MAGNITUDE DIAGRAM #68: Diagram of the magnitudes of stars versus their color index.

COMA #40: The gaseous envelope surrounding a comet.

COMET #40, 56: A "dirty ice ball" orbiting a star.

CONSERVATION OF ANGULAR MOMENTUM #56, 95: A principle of physics that is responsible for bodies speeding up in their spin as they contract. The equation is $L = m(v)r \sin\Theta$

CONSTELLATION #1, 2: An outline made by connecting the stars in the sky, usually a figure from mythology.

CONTACT #7: A science fiction novel and movie about the search for extraterrestrial life featuring a contact signal from the star Vega.

COPERNICUS, NICOLAUS #102: 16th Century astronomer and creator of the Heliocentric model of the universe.

CORONA #58: The outer atmosphere of the Sun; the part of the Sun visible during a total solar eclipse.

COSMIC YEAR #60: One revolution of the Sun around the Milky Way Galaxy, taking about 225 million Earth years.

COULOMB BARRIER #82: The electromagnetic force resulting from a particular charge on a like charge.

CRAB NEBULA (Ml) #8, 91: A remnant of the supernova of July 4, 1054.

CYGNUS #7, 70: The northern hemisphere constellation of the Swan.

CYGNUS X-1 #7: The first black hole candidate, located in Cygnus.

DARK ENERGY #71: A mysterious "anti-gravity" force that seems to be accelerating the expansion of the universe.

DARK MATTER #74: Non-baryonic matter in space that cannot be seen but is detected due to its gravitational effects.

DAUGHTER NUCLIDE #35: Decay products of nuclear reactions.

DEFERENT #101: In Ptolemaic theory, a wheel that carries a planet around in a perfect circle.

DEGENERATE MATTER #99: Super-dense matter, usually with a density greater than 10^6 grams/cm^3.

DELAY IN MOONRISE #21: The amount of time the Moon will rise later today than it did yesterday.

DELPHINUS #7: Northern hemisphere constellation lying beside the Summer Triangle.

DENEB #7: Star of magnitude 1.4 at the tail of Cygnus the Swan.

DE REVOLUTIONIBUS ORBIUM COELESTIUM #102: The title of Copernicus' book, *On the Revolutions of the Heavenly Spheres.*

DESERTIFICATION #37: Changing fertile land into a desert.

DIANA #6: Goddess of the Moon.

DIURNAL CIRCLE #1: Path of a star on the Celestial Sphere.

DIURNAL MOTION #28: The apparent daily movement of celestial bodies across the sky due to Earth's rotation.

DODECAHEDRON #101: Twelve sided solid with each side being a regular pentagon.

DOGON #3: Native tribe from the central plateau of Mali, south of the Niger River.

DOPPLER EFFECT, THE #61: A change in wavelength or frequency of a wave due to the source of the wave either approaching or receding from the observer. Its utilization allows the radial velocity (velocity along line of sight) of an object to be determined.

DRAKE EQUATION #96, 97: Equation for the number of communicative civilizations extant in the galaxy at this time.

DUBHE #1: Star of magnitude 2 in the bowl of Ursa Major.

DUMBBELL NEBULA #87: A fine planetary nebula (M27) within the constellation Vulpecula.

DUST TAIL #40: The tail of a comet made of dust from the nucleus and driven into space by solar radiation pressure.

$E = MC^2$ #80: The equation for the energy accompanying the mass laws in a nuclear reaction.

"EAGLE, THE" #30: The lunar excursion module on the Apollo XI mission.

ECLIPSE PATH (ALSO, "PATH OF TOTALITY") #42: Path from which you can see a total solar eclipse while standing on the Earth.

ECLIPSE SEASON #15: Time during the year when the line of nodes (places where the Moon's orbit around the Earth intersect the ecliptic) points to the Sun.

ECLIPSE WINDS #42: Winds that occur before and during a solar eclipse because of the pressure difference between the shadow and where the Sun is shining.

ECLIPSING BINARY STAR #4: A type of binary star in which the companion stars move in front of each other, from an Earth observer's point of view, as they orbit.

ECLIPTIC #31, 34, 66: Apparent path of the Sun through the sky during the year. Also, the Earth-Sun plane.

EINSTEIN, ALBERT #38, 77, 78, 88: Physicist responsible for the Special and General Theories of Relativity.

ELECTROMAGNETIC RADIATION #38, 65, 95, 98: Energy field that includes radio, microwave, infrared, visible, ultraviolet, x-rays and gamma rays.

ELONGATION ANGLE #46: Apparent angle between a planet and the Sun.

ENERGY #80: The ability to do work.

ENERGY OF POSITION #80: The amount of energy a body has by virtue of its position in a field.

EPICYCLE #101: A small circle which carries a planet around the deferent in the Ptolemaic model of the universe.

EQUATION OF TIME #66: The difference between apparent solar time and mean solar time on a given date.

EQUATOR OF THE EARTH #18: A line halfway between the poles on the surface of the Earth.

ERATOSTHENES #101: Director of the Great Library of Alexandria who measured the circumference of Earth to an accuracy of 1 percent.

ERIDANUS #8: An equatorial constellation of the Po River.

EXOPLANET (ALSO, EXTRA-SOLAR PLANET) #61: A planet orbiting a star in a system outside the Solar System.

FAR SIDE OF MOON #12: The side of the Moon that cannot be seen from Earth's point of view.

FIRST CONTACT #42: When the Moon just intrudes between the Sun and the Earth in a total solar eclipse.

FIRST QUARTER MOON #16: After the New Moon, when the Moon is at quadrature, or 90 degrees from both the Earth and Sun.

FISSION #20: When a body splits apart, usually due to rapid rotation. Also, a nuclear reaction wherein an atomic nucleus splits into two or more components.

"FLASH" SPECTRUM #58: The spectrum of the solar chromosphere.

FORBIDDEN PLANET #7: 1950s science fiction movie.

FOSSIL FUELS #37: Compounds in the Earth's crust made from fossilized remains of plants.

FULL MOON #16: When the Moon is at opposition to the Sun.

FUSION #38: A nuclear reaction in which relatively small atomic nuclei are fused together into bigger ones.

FUSION SHELL #87: In stars going from main sequence to giant, a zone of fusion which stabilizes the star or causes it to expand.

FUSION TEMPERATURES #33: The temperature at which fusion takes place, which is a function of the pressure of the gas; it is about 13 million Kelvin for hydrogen fusion.

GALACTIC CENTER #67: The center of the galaxy; it is 10 to 12 kiloparsecs away from us, in the direction of Sagittarius.

GALACTIC HALO #105: The region above and below the galactic plane in which we find Population II objects.

GALILEAN TRANSFORMATION #77: A set of equations which transform a function from one coordinate frame to another which is in uniform motion with respect to the first frame.

GALILEI, GALILEO #102, 103: 17th Century astronomer who ran afoul of the Italian Inquisition. He built an early telescope with which he observed the Sun, Moon, and planets, and made great discoveries.

GALLE, JOHANN FRIEDRICH #53: 19th Century astronomer who helped in the discovery of Neptune.

GAMMA RAY #49, 80, 81, 89: A photon of electromagnetic radiation that has a wavelength less than 1 angstrom.

GAS (ION) TAIL #40: One of the tails of a comet formed by the solar wind. The other is the dust tail.

GAUSS, FREDERICK #41: Director of the observatory in Gottingen, Germany, in 1807. His mathematical method was used to reacquire Ceres after it had been initially lost by Piazzi.

GEMINIDS #22: The meteor shower that occurs around December 13th at a rate of about 50 an hour.

GENERAL THEORY OF RELATIVITY #78: Albert Einstein's theory of the concept of gravity that posits the warping of space and time by masses.

GIANT STAR #86: A star with a diameter of up to three hundred solar diameters.

GIBBOUS MOON #16: The lunar phases that occur between first quarter and full, and between full and last quarter.

GLOBAL WARMING #37: The theory that the troposphere is warming due to the introduction into the atmosphere of anthropogenic heat-absorbing gases.

GLOBULAR CLUSTER #92, 94, 105: One of about 200 star clusters that orbit the halo of the galaxy from 3 to 150 kiloparsecs in long elliptical paths.

GOLDEN FLEECE #7: The golden fleece of Aries the Ram, object of the quest of Jason and the Argonauts.

GRAVITATIONAL RADIATION #65: In the General Theory of Relativity, the oscillation of the spacetime medium brought about by an oscillation of a quantity of mass.

GRAVITATIONAL REDSHIFT #78: A redshift brought about by slowing in a gravitational field.

GRAVITY #56, 77: A force of attraction proportional to the product of the masses of objects and inversely proportional to the square of the distance between their centers.

GRAVITY OF A PLANET #44: A quantity which is proportional to the mass of the planet and inversely proportional to the square of its radius.

GREAT RIFT, THE #67, 70: A lane of cosmic dust lying parallel to Cygnus and parallel to the plane of the Milky Way.

GREENHOUSE EFFECT #19, 37: The trapping of infrared radiation by certain gases or translucent material.

GREENHOUSE GASES #37: Gases that trap infrared radiation. Examples are water vapor, carbon dioxide, and methane.

GUARDIANS OF THE POLE #10: The two stars, Kochab and Pherkad, at the end of the dipper bowl in Ursa Minor.

GUTH, ALAN #71: Theoretical physicist and cosmologist who proposed the idea of inflation of the very early universe.

GYROSCOPIC ACTION #1: Angular momentum created by an object spinning on an axis.

HALLEY, EDMOND (EDMUND) #104: 17th Century English astronomer who urged Newton to share his discoveries.

HARVEST MOON #31: The September full Moon, said to aid farmers in harvesting crops by providing light after sunset.

HAWKING, STEPHEN #90: 20th and 21st Century British theoretical physicist who has explained black holes and General Relativity. His best known book is *A Brief History of Time.*

HAWKING RADIATION #90: Stephen Hawking's hypothesis that black holes leak radiation.

HELIOPAUSE #58: The region where the solar wind merges with the general magnetic field of the galaxy.

HERA #4, 70: Greek goddess of women and marriage. The wife and sister of Zeus, daughter of Rhea and Kronos.

HERCULES CLUSTER #94: A globular cluster of about a million stars lying in the Hercules constellation.

HERMES #2, 7, 46: Messenger of the Greek gods, also the god of speed; his Roman name was Mercury, which is the planet with the fastest annual orbit.

HERSCHEL, FRIEDRICH WILHELM #51, 77: 18th Century British astronomer, known as Sir William Herschel, who discovered Uranus.

HODGES, ANN #24: Alabama woman struck by a 9 pound meteorite that crashed through her roof.

HOOKE, ROBERT #104: English natural philosopher and adversary of Isaac Newton; known as the "Father of Microscopy" for his work viewing cells; an early advocate of Darwin's theory of natural selection.

HOUR CIRCLES #28: The daily paths made by stars, described as if the Earth were stationary and all other bodies were moving around it, drawn on the Celestial Sphere.

HUBBLE, EDWIN #63, 69, 105: American astronomer who discovered the existence of galaxies beyond the Milky Way.

HUMASON, MILTON #69: Mule-team driver who helped build the Mt. Wilson observatory; his observational skills allowed him to contribute to Hubble's project of ascertaining the radial velocities of the galaxies.

HUNTER'S MOON #16, 31: The October full Moon.

HYDROGEN #44: The lightest and most abundant element in the universe. It is highly flammable and has the atomic number one.

HYDROGEN FUSION #32, 81, 86: A nuclear reaction in which hydrogen atoms are fused together to produce helium atoms.

IBT AL JAUZAH #5: Betelgeuze, Arabic name for one of the stars of the constellation Orion, meaning "the armpit of the central one."

INFRARED #19: Light of a wavelength of 750 nanometers to 1 millimeter.

INNER PLANETS #56: The planets Mercury, Venus, Earth, and sometimes Mars.

INTERNATIONAL DATE LINE #66: The 180 degree meridian on Earth's surface. A traveler subtracts one day going east and adds one day going west when crossing it.

IONIZED HELIUM ATOMS #58: Atoms with a charge either + or − . It could be plus plus or minus minus. Plus plus means two electrons are missing. Minus minus means there are two additional electrons.

ION (GAS) TAIL #40: One of the two tails of a comet. It occurs because of the solar wind.

IRRATIONAL NUMBERS #101: Numbers that are not expressed as fractions or ratios of whole numbers.

JASON #7: A mythical person who led the Argonauts in seeking the Golden Fleece.

JUNO #9: Roman name for the Greek goddess Hera, queen of the goddesses and wife and sister to Zeus.

JUPITER #44: Fifth and largest planet of the Solar System.

KELVIN #38: The unit of absolute temperature.

KEPLER, JOHANNES #102, 103: 17th Century astronomer responsible for the three laws of planetary motion.

KEPLERIAN ROTATION #74: When the values of successive velocities in a system reduce over distance (the farther a planet is from the Sun, the slower is its orbital motion).

KEPLER'S LAWS #103: Three laws of planetary motion developed by the astronomer.

KEPLER'S THIRD LAW #93, 103: Kepler's law that says that the mass of the system is equal to the separation cubed, divided by the period squared.

KERR, ROY # 90: Mathematician who in 1963 discovered a way to solve the double valued equation for a rotating black hole.

KOCHAB #10: One of the two (with Pherkad) Guardians of the Pole.

KUIPER BELT #57: Extending from the planet Neptune to about 50 AU, this region is the home of comets and at least three dwarf planets (and possibly up to 200).

LAW OF GRAVITY #104: According to Newton, the force all material objects exert on each other.

LAWS OF MOTION #77, 104: Newton's explanation of the relationships of the forces acting on objects and the motion of the objects.

LEAVITT, HENRIETTA #84: Early 20th Century American astronomer whose observations of Cepheid variable stars led to an accurate means of measuring distances in the universe.

LEMÂITRE, GEORGES #64, 69: Belgian priest and astronomer who in 1931 used Hubble's work to theorize the Big Bang.

LEONID SHOWER (LEONIDS) #22: The meteor shower that occurs around November 16th at a rate of about 15 an hour.

LE VERRIER, URBAIN #53, 77: 19th Century French astronomer who correctly used the apparent erratic behavior of Uranus to posit the existence and location of Neptune.

LGM (LITTLE GREEN MEN) #95: Cambridge University student Jocelyn Bell's whimsical name for unexplained radio pulses received from space while studying quasars and neutron stars. Later named pulsars.

LIGHT YEAR #2: A measure of distance covered by a beam of light in one Earth year. Equivalent to approximately 6 trillion miles.

LIQUID HYDROGEN #44: The substance of which the outer planets in the Solar System mostly consist.

LITTLE DIPPER #1: An asterism within the constellation Ursa Minor.

LOBACHEVSKIAN GEOMETRY #75: A replacement geometry for Euclidean, in which the sum of the angles in a triangle is less than 180 degrees.

LOCAL HOUR ANGLE #66: The angle between an object on a meridian and the sigma point on the Celestial Equator. It equals the time elapsed since the object was last on the meridian.

LOCAL MEAN TIME #66: The hour angle of the mean Sun plus twelve hours; its calculation is necessitated by Earth's elliptical orbit causing the apparent speeding up and slowing down of the Sun.

LOCAL MERIDIAN #66: Longitudinal circle around the Earth on which the observer is located.

LOCAL ZENITH #66: The point directly above an observer on the Earth's surface.

LOSS OF BIODIVERSITY #37: The loss of distinct biological species and ecosystems signalling a decline in the overall health of the biosphere.

LOST PLEIADE #8: The faintest of the seven stars that make up the Pleiades.

LOWELL, PERCIVAL #39: Astronomer who predicted the location of a ninth planet in 1880, later discovered by Clyde Tombaugh in 1930. This planet became known as Pluto after 11 year old Venetia Burney suggested the name.

LUMPY 3K BACKGROUND #83: The result of gravitational attraction of celestial bodies into groupings of different sizes, thus producing a universe lacking a smooth consistency.

LYRA #7, 49, 87: Small constellation that contributes stars to the Summer Triangle and includes Vega, one of the brightest stars in the sky.

M #8: The letter designating Messier objects.

MAIN SEQUENCE STAR #33, 81, 82: What is created when hydrogen fusion temperatures are reached within a protostar.

MAJOR PLANETS #56: In the Solar System, the eight planets from Mercury to Neptune. Also, defined as spherical bodies which are orbiting a star and have gravitationally cleared their orbit of other bodies.

MARINER 9 #45: First spacecraft to photograph water channels on Mars, 1970 to 1972.

MARS #45: Fourth planet from the Sun.

MASS #44: The measurement of a body determined by its gravity.

MASS LOSS #82: The elements heavier than helium ejected during the dying of a star.

MAXWELL'S EQUATIONS #77: The electromagnetic equivalent of Newton's laws describing mechanical phenomena.

MEAN SUN #66: For the purpose of measuring time, a uniform speed astronomers assign to the Sun as it circles the Celestial Equator.

MERAK #1: One of two stars in the bowl of the Big Dipper that points the way to the North Star.

MERCURY #46, 77: First planet from the Sun, with the fastest annual orbit. In Greek mythology, Hermes, the speedy messenger of the gods.

MEROPE #6: In classical mythology, the woman Orion fell in love with and raped when she was withheld from him.

MESSIER'S CATALOGUE #8, 87: 110 astronomical objects catalogued by the French astronomer, often observable without extreme magnification.

METEOR #22, 24: The flash of light made by a meteoroid as it burns up in the Earth's atmosphere.

METEOR CRATER #24: The depression in Northern Arizona left by the impact of a meteoroid/asteroid.

METEORITE #24: A rock from space after it has landed on Earth.

METEOROID #24: Originating from the Asteroid Belt, a chunk of rock/metal from space, less than 150 feet in diameter, while it is still in space (or traveling through Earth's atmosphere).

METEOR SHOWER #22: More than the seven per hour that is usual constitutes a shower; showers are named for the radiant points from which they seem to originate. They occur due to Earth intersecting, and colliding with, the small meteoroids left in a comet's orbit.

MICHELSON-MORLEY EXPERIMENT #77: In Special Relativity theory, the experiment that produced the discovery that the speed of light did not change as the speed of the observer changed.

MID-ATLANTIC RIDGE #17: A place under the Atlantic Ocean where magma comes from the mantle and spreads east and west.

MIDNIGHT SUN #23, 25: The tilt of the Earth's axis causing certain areas near the poles to have constant sunshine for months at a time.

MILKY WAY GALAXY #67, 70, 93, 94, 105: Our home galaxy consisting of 200 to 400 billion stars.

MINI BLACK HOLE #76: A theoretical black hole less than 3.0 solar masses (i.e., has the mass of a mountain but the size of a grain of sand).

MINKOWSKI, HERMANN #88: German mathematician who proposed the concept of spacetime.

MINOR PLANETS #56: The millions of small, non-spherical planets found mostly in the Asteroid Belt between Mars and Jupiter.

MIRACLE YEAR #104: 1664, the year Newton developed the laws of gravity and motion as well as other important mathematical principles.

MIRANDA #51: One of Uranus' 27 satellites, uniquely formed of several materials.

MISSING MASS #74: The dark matter in clusters of galaxies.

MOLINA, M. J. #14: Chemist who, with F.S. Roland, showed that the Earth's ozone layer was being depleted.

MOON ILLUSION #13: The appearance of the Moon just above the horizon seeming larger than when seen directly overhead.

"N" #96, 97, 98: In the Drake Equation, the number of communicative civilizations possibly extant in the galaxy at this time.

NEBULA #63: A cloud of gas in space.

NEPTUNE #44, 53, 77: The eighth planet from the Sun, discovered mathematically by computing residuals from Uranus.

NEUTRINO #80, 89, 91: A particle which travels near the speed of light and is extremely non-interactive.

NEUTRON STAR #65, 95: A stellar core of between 1.4 and 3.0 solar masses, when gravity drives the electrons inside the protons, forming an object that is a mass of neutrons.

NEW MOON #16, 42: When the Moon is in the same direction as the Sun, rendering it completely unlit on the side facing the Earth.

NEWTON, ISAAC #77, 104: 17th Century English scientist who developed laws of gravity and motion as well as other important mathematical principles.

NEWTON'S LAWS #104: Isaac Newton's laws of gravity, motion, and spectroscopy.

NORTH AMERICAN NEBULA #7: An emission nebula within the constellation Cygnus.

NORTH AMERICAN PLATE #17: Earth's crustal plate that contains all of North America except part of California.

NORTH CELESTIAL POLE #28, 66: On the Celestial Sphere, the point directly above the Earth's north pole.

NORTH POINT #28: Intersection of the meridian and the horizon.

NORTH POLE #28: Intersection of the Earth's axis and the meridian.

NUCLEOSYNTHESIS #82, 85: The creation of elements heavier than hydrogen.

NUCLIDES #35: Decay products of nuclear reactions.

OBERON #51: One of Uranus' 27 satellites.

OCCAM'S RAZOR #3, 11: A principle that states that the simplest explanation is usually the correct one.

OENOPION #6: The king who blinded Orion after he assaulted his daughter.

OORT CLOUD #56: A spherical cloud of comets possibly 50,000 AUs from the Sun; the farthest region in the Solar System.

OPEN (GALACTIC) STAR CLUSTERS #94, 105: Groupings of ten to a few thousand of stars with no real symmetry.

ORBIT #55: The curved path made by one object moving around another, held in place by gravity.

ORIONIDS #22: The meteor shower that occurs around October 21st at a rate of about 20 an hour.

OSIRIS #8: One of the oldest Egyptian gods, associated with myths of the underworld and the dead.

OUTER PLANETS #56: The Solar System's planets Jupiter, Saturn, Uranus, Neptune, and sometimes Mars.

OWL NEBULA #87: A planetary nebula in Ursa Major.

OZONE DEPLETION #14: The reduction of the Earth's protective layer of ozone by chlorofluorocarbons, and halogenated ozone depleting substances.

PACIFIC PLATE #17: Earth's crustal plate that contains part of California and all of the Pacific Ocean.

PARENT NUCLIDE #35: The first product of nuclear decay; linked to the daughter nuclide.

PARSEC #2: Using the method of stellar triangulation from Earth, the distance to a hypothetical object if that object's PARallax equals exactly one arc SECond. Also, a distance equivalent to 3.26 light years.

PARTIAL SOLAR ECLIPSE #42: When the Sun is "partially" blocked by the Moon.

PENUMBRA #42, 59: Partial shadow cast by the Moon onto the Earth during an eclipse. It surrounds the umbral (total) shadow. Also, the lighter gray perimeter of a sunspot.

PENZIAS, ARNO #64: One of two physicists (with Wilson) who accidentally discovered evidence of the Big Bang when recording the 3 Kelvin microwave background radiation of the universe.

PERIHELION #77: The closest point to the Sun in a planet's orbit.

PERILUNE #30: The closest point to the Moon in a satellite's orbit around it.

PERIOD #4, 12, 84: The length of time it takes to complete a full cycle, be it rotation, revolution, light output, etc.

PERIOD-LUMINOSITY RELATION #84: A graph of the period versus the luminosity of Cepheid variables.

PERSEIDS #22: Usually the most prolific meteor shower of the year that occurs around August 12 at a rate of about 70 an hour.

PERSON IN THE MOON #36: In various mythologies, the imaginary being said to reside or appear on the Moon.

PHASES OF THE MOON #16: Beginning with "new Moon" and ending with "new Moon" one cycle later, the changing appearance of the Moon on its 30 day orbit around the Earth due to an observer seeing different portions of its illuminated side.

PHERKAD #10: One of the two (with Kochab) Guardians of the Pole.

PHOTOSPHERE #58, 59: The apparent surface of the Sun, from which its light is emitted.

PIAZZI, GUISEPPE #41: Sicilian astronomer who discovered Ceres in 1801.

PIEZOELECTRIC CRYSTALS #65: Crystals that are able to generate electrical impulses in response to pressure.

PLANETARY NEBULAE #33, 49, 86: What remain after stars have collapsed and pushed out shells with blue-green appearances, similar to the planets Uranus and Neptune, after which these nebulae were named.

PLANET X #57: Any undiscovered planet in our Solar System beyond the orbit of Uranus.

PLATE TECTONICS #17: The converging and diverging movement of the Earth's crust in large zonal divisions.

PLATO #101, 102, 103: Greek philosopher who devised erroneous geometric models as keys to the structure of the universe.

PLEIADES #8, 94: The Seven Sisters, a cluster of stars located in the shoulder of the constellation Taurus. Also known as "The Seven Goats," "The Six Wives who ate Onions" and "Subaru" (in Japanese).

PLUTO #57: Once the ninth major planet in the Solar System, now demoted to dwarf planet status.

POINTER STARS #10: Merak and Dubhe in the Big Dipper. When connected by a line, these are used by astronomers to locate the North Star.

POLARIS #1, 10, 28: The North Star, located near the North Celestial Pole.

POPULATION I OBJECTS #105: Galactic clusters of objects born out of materials left by the death of the first (Population II) objects. These younger stars have higher metal content.

POPULATION II OBJECTS #105: Globular clusters, the oldest observed stars that formed in the galaxies.

PRECESSION #1, 77: The wobbling of a celestial body on its axis as it rotates, caused by gyroscopic action.

PRECESSION OF MERCURY #77: The twisting of Mercury's orbit brought about by the relativistic effect on time.

PRECESSION OF THE EQUINOXES, THE #10: The periodic wobbling of Earth's axis of rotation brought about by the gravitational action of the Moon and planets on its equatorial bulge. The Earth completes one period of precession in approximately 26,000 years.

PRIME MERIDIAN #66: On Earth, the line (meridian) of longitude extending from the north pole to the south pole which intersects Greenwich, England. Where this meridian intersects the equator is the 0 degree point of longitude.

PRINCIPIA, THE (THE MATHEMATICAL PRINCIPLES OF NATURAL PHILOSPHY) #104: Isaac Newton's seminal three volume treatise published in 1687, considered one of the most important works of science ever written.

PRINCIPLE OF EQUIVALENCE #78: The concept that all inertial effects are indistinguishable from gravitational effects.

PROJECT OZMA #98: American astronomer Frank Drake's project that sent radio waves into space during the first SETI (Search for Extraterrestrial Intelligence).

PROTON-PROTON REACTION #49: The fusion process that powers the Sun by transforming hydrogen into helium.

PROXIMA CENTAURI #61: The closest star to the Solar System at 4.24 light years away. One of the 3 stars in the Alpha Centauri star system.

PTOLEMY #3, 34, 102: Claudius Ptolemaeus, Greek astrologer who devised an early scientific model of the universe.

PTOLEMAIC GEOCENTRIC UNIVERSE #102: A system of "epicycles" and "deferents" that imperfectly predicted planetary motion.

PULSAR #95: A rotating neutron star only visible when beaming or pulsing its electromagnetic radiation in a manner described as the lighthouse effect.

PYTHAGOREANS #101: The earliest scientists, circa 600 BCE, who believed that nature's happenings were explainable by human intellect.

QSO #100: Quasi-stellar object, or quasar.

QUARKS #82: The particles that comprise protons and neutrons.

QUASARS #95, 100: Star-like objects emitting radio waves; they turned out to be distant galaxies seen in the process of formation.

R AND S PROCESSES #85, 86: Rapid and slow neutron capture, at work when forming elements heavier than hydrogen and helium in the atmosphere of a giant and supergiant star.

RADIAL VELOCITY #69: Velocity in the line of sight.

RADIANT POINT #22: The direction in space from which meteors appear to originate.

RADIATION PRESSURE #40, 68, 81: The force supplied by a photon when it collides with a particle.

RADIO ASTRONOMERS #99: Those who study celestial bodies at radio wavelengths.

RED GIANT #33, 49, 86: A dying star of 3 to 500 solar diameters.

RED MOON #16: The appearance of the full Moon when seen through Earth's atmosphere near the horizon.

REDSHIFT #69: A shift of an entire spectrum of an object toward the red due to the recession of the object.

REST MASS ENERGY #80: In Relativity theory, the energy inherent in mass when it is at rest.

RETRO-ROCKETS #20: Thrusting engines used to decelerate a spacecraft by firing in the direction of travel.

RIEMANNIAN GEOMETRY #75: Positive form of non-Euclidian geometry in which the sum of the angles in a triangle is greater than 180 degrees.

RIGEL #5: Star in the constellation Orion that indicates the left leg.

RIJL JAUZAH AL YUSRA #5: In the constellation Orion, the Arabic name for the "left leg of Jauzah."

RING NEBULA #49, 87: Example of a planetary nebula in Lyra; a gas shell surrounding the star in the last stages of its life.

RISING SEAS #37: An after-effect of global warming caused by the melting of the polar ice caps.

ROLAND, F.S. #14: Chemist who, with M.J. Molina, showed that the Earth's ozone layer was being depleted.

RR LYRAE #105: Pulsating variable stars with short periods between pulsations.

RUBIN, VERA #74: Astronomer whose work in measuring the speed of globular clusters led to conclusions about the existence of dark matter as an explanation for the unanticipated greater mass of the galaxy.

RUNAWAY GREENHOUSE EFFECT #52: Too much of a good thing: The result of human and other activity producing too much greenhouse gases, leading to global warming.

S PROCESS #82: The process of creating elements through slow neutron capture.

SAGITTA #7: The Arrow, a minor constellation within the Summer Triangle.

SAGITTARIUS #67: Constellation of the Archer in the direction of the center of the Milky Way.

SAIPH #5, 6: The star that indicates the right leg of Orion in the constellation.

SATELLITES #56: Smaller bodies that orbit planets.

SATURN #44, 48: Sixth planet from the Sun, a liquid-gas giant surrounded by ice rings.

SATURN'S RINGS #48: An infinity of small ice particles all in their own separate orbits around Saturn.

SCHWABE, HEINRICH #59: German astronomer who observed that the number of sunspots varied with time.

SCORPIUS #9: Constellation that contains Antares, a red supergiant.

SECOND CONTACT #42: When totality is achieved during a total solar eclipse.

SETI #98: Acronym for "Search for Extraterrestrial Intelligence," a project launched by astronomer Frank Drake in 1960.

SHADOW BANDS #42: The unexplained effect of thousands of bands of light and shadow moving across the landscape during totality of a solar eclipse.

SHAPLEY, HARLOW #92: American astronomer who mapped the Milky Way and established the distances of globular clusters. He correctly calculated the Sun's location within our galaxy.

SHELL FUSION #33: The process in which dying stars undergo fusion in a "shell" of gas around their cores as these cores collapse and increase in temperature.

SHURNARKABTISHASHUTU #8: A star in the constellation Taurus; Babylonian for "the end star of the southern horn of the bull," the longest name in the sky.

SIDEREAL DAY #66: The period of rotation of the Earth with respect to the stars.

SIDEREAL TIME #66: Using a star as a reference point to keep track of the passage of time.

SIGMA POINT #66: In telling sidereal time, the reference point where the local meridian crosses the Celestial Equator.

SIRIUS #3, 62: Located in Canis Major, the brightest star in the night sky.

SIRIUS B #3, 99: A white dwarf companion star that orbits Sirius every 50 years.

SLIPHER, V. M. #69: Edwin Hubble contemporary who discovered that galaxies were redshifted.

SLOW NEUTRON CAPTURE #82, 86: During nucleosynthesis, when a lighter element takes on a neutron and raises the atomic number by one, thus creating a heavier element necessary in planet formation.

SOLAR CORONA #58: The superheated crown surrounding the Sun's photosphere, where the solar wind is generated.

SOLAR ECLIPSE #42: The occurrence where the Moon blocks the disk of the Sun.

SOLAR MAGNETIC STORMS #58: The Sun's energy transmitted by solar winds throughout the Solar System.

SOLAR SYSTEM #2, 56: The Sun and the bodies which orbit it.

SOLAR WIND, THE #40, 58: The result of the Sun's corona expanding, sending high velocity protons and electrons throughout the Solar System.

SOLID WHEEL ROTATION #74: Rotation that exhibits increasing velocities for bodies located farther away from the center, as if attached to the spoke of a wheel (as opposed to the slowing orbital velocities found in Keplerian rotation).

SOLSTICE #25, 50: The two times (December 21 or 22 and June 21) every year when the Sun seems to stand still, ceasing its southward and northward movements in the sky and reversing directions. Latin for "stationary Sun."

SOUTH POINT #66: Position on the Celestial Sphere held by the farthest point south on the horizon.

SOUTH POLE #23, 28: Southernmost point on Earth whereupon the axis is located.

SPACETIME #65, 88: Location in both space and time. Addition of the fourth dimension of time to describe an object's location characterizes the universe as a spacetime continuum.

SPECIAL THEORY OF RELATIVITY #77, 80: Albert Einstein's 1905 theory that the speed of light in free space is the same for all observers regardless of their motion relative to the light source. It contradicts the notion that intervals of time between events are equal for all observers.

SPECTRAL TYPE #5: A classification system that rates stars by their temperature class.

SPEED OF LIGHT #64, 65, 78: Denoted by the letter c, 186,000 miles per second or 300,000 kilometers per second.

SPICA #2: The brightest star in the constellation Virgo.

SPIRAL GALAXY #70: Galaxy characterized by pinwheel-like arms that extend outward from the center.

SPREADING CENTER #17: New crust formation resulting from molten material coming up from the Earth's mantle.

STEADY STATE #64: Competing theory to the Big Bang, positing that the universe is and has been the same at all times and in all places.

STELLAR MASS BLACK HOLE #76, 90: A black hole with a mass between 3 and approximately 100 solar masses, created from the collapse of the core of a supergiant star.

STRONG NUCLEAR FORCE #95: The strongest known force in the universe, located within an atomic nucleus.

SUBDUCTION ZONE #17: In plate tectonics, where two plates come together.

SUMMER SOLSTICE #25: June 21 or 22, the point in the sky where the Sun reaches its farthest northern position.

SUMMER TRIANGLE #1, 7: Asterism made up of three of the brightest stars in the summer sky, Deneb, Vega, and Altair.

SUNSPOTS #58, 59: Magnetic storms generated in the photosphere of the Sun. They are slightly cooler and thus, appear darker than the surrounding region.

SUPERGIANT #9, 86: A star with sufficient mass to eventually achieve carbon fusion and successive fusions to create heavier elements.

SUPER MASSIVE BLACK HOLE #76: Black hole with up to tens of billions solar masses residing at the center of a galaxy.

SUPERNOVA #68, 86, 91: A stellar explosion caused by gravitational collapse.

SUPERNOVA 1987A (SN1987A) #91: A supernova which was observed in the Large Cloud of Magellan. Astronomers were made aware of it by the detection of escaping neutrinos.

SYENE #101: Now Aswan, in Egypt. Place where Eratosthenes compared the angle of the Sun's shadow on the solstice to the one in Alexandria to calculate the circumference of the Earth.

SYLACAUGA #24: Alabama town where a meteoroid crashed through a house in 1954.

TAURUS #8: The constellation of the Bull.

TEMPERATURE CLASS #5: Same as spectral type, for stellar classification purposes.

TERRESTRIAL PLANETS #82: Planets with solid rock/metal surfaces. In the Solar System, Mercury, Venus, Earth, and Mars.

THIRD (LAST) QUARTER MOON #16: When the Moon has traversed three quarters of the way in its orbit around the Earth starting from new Moon.

THREE KELVIN (3K) MICROWAVE BACKGROUND #64, 83: The background radiation (at 3 Kelvin) of the universe, consistent with what astronomers expected from the Big Bang theory.

TITANIA #51: One of the satellites of Uranus.

TOTAL LUNAR ECLIPSE #15: When the Moon entirely enters the Earth's umbral shadow, turning a coppery-red.

TOTAL SOLAR ECLIPSE #42: When the Moon totally blocks the disk of the Sun.

TRIPLE ALPHA PROCESS #85, 86: Heavy element fusion accomplished by the process of three helium nuclei being transformed into carbon.

TROPIC OF CANCER #34: The +23.5 degree line of latitude on Earth indicating the northern limit of the Sun's path.

TROPIC OF CAPRICORN #34: The −23.5 degree line of latitude on Earth indicating the southern limit of the Sun's path.

TWIN PARADOX #73, 78: The case in Special Relativity where the twin traveling and returning at a very high velocity ages less than the stay-at-home twin.

UFOLOGY #98: The study of Unidentified Flying Objects, not yet considered a science.

UMBRA #42, 59: The dark central region of a shadow. A region of total shadow.

UMBRIEL #51: A satellite of Uranus.

URANUS #44, 51, 53, 77: Seventh planet from the Sun.

URSA MAJOR #1, 10, 87: The Big Bear constellation, contains the Big Dipper.

URSA MINOR #1, 10: The Little Bear constellation, contains the Little Dipper.

VEGA #7: One of the corner stars in the Summer Triangle. Brightest star in Lyra.

VENUS #52: Second planet from the Sun.

VERNAL EQUINOX #18, 66: March 21, the point in the sky where the Sun crosses the Celestial Equator midway between winter and summer.

VISIBLE LIGHT #19: Light which can be seen by the naked eye.

VULCAN #77: Name given to the new planet that astronomer believed existed because of perturbations in Mercury's orbit, later explained by Relativity theory.

VULPECULA #7, 87: The Fox constellation within the Summer Triangle.

W VIRGINIS STARS #105: Pulsating variable Type 2 Cepheids found within the halo of the galaxy.

WANING GIBBOUS MOON #16: After the full Moon, the 7.3 days of lunar phases before 3rd (last) quarter Moon.

WATER HOLE #98: The frequency of radiowave transmission at which the galaxy transmits, so-called because it is located in a region or hole between the frequency of water and hydroxide.

WATER VAPOR #19: A greenhouse gas that warms the Earth, keeping temperatures from getting too cold for life.

WEBER, JOSEPH #65: Astronomer who first set up a gravity wave detector.

WHITE DWARF #3, 87, 99: The core of a giant star after the planetary nebula phase.

WHITE HOLES #90: The theoretical other side of a black hole that expands as the black hole collapses inward.

WILSON, ROBERT #64: One of two physicists (the other being Arno Penzias) who inadvertently discovered the 3-K Microwave background while experimenting with communication techniques.

WODEN #46: The early German name for Mercury, god of the sky.

WORLDLINE #88: An interval of spacetime, the unique path of travel of an object tracked through four dimensions.

WORM HOLES #90: Also "Einstein-Rosen Bridge," the theoretical passageways through black holes (and hyperspace) to another universe or another place in ours.

ZONE TIME #66: On Earth, regions 15 degrees wide that by convention share the same mean solar time.

ZWICKY, FRITZ #74: Astronomer whose observations of galaxy clusters contributed to the conclusion that mass is missing or hidden from observers.

APPENDICES

Major Planet Data

Planet	Dia. (Earth Dia.)	Mass (Earth Mass)	Density (g/cm3)	Rotation Rate	Revolution Period	Axial Tilt (degrees)	Orbital Eccentricity
Mercury	0.38	0.055	5.44	59d	88d	0	0.206
Venus	0.95	0.815	5.24	243d	225d	177 (or 3)	0.007
Earth	1.00	1	5.52	23h 56m	365.24d (1 y)	23.5	0.017
Mars	0.53	0.108	3.94	24h 37m	1.9y	24	0.093
Jupiter	11	317.8	1.34	9h 50m	11.9y	3.1	0.048
Saturn	9.5	95.15	0.69	10h 14m	29.5y	26.4	0.056
Uranus	4.0	14.54	1.27	17h 14m	84y	98 (or 82)	0.046
Neptune	3.9	17.15	1.66	16h 3m	164.8y	29	0.010

20 Brightest Stars As Visible From Earth

Name	Distance (LY)	m	M	Spectral Type/Lum. Class
1. Sirius	8.6	-1.46	1.4	A1V
2. Canopus	74	-0.72	-2.5	F0II
3. Alpha Centauri	4.3	-0.27	4.4	G2V + K1V + M5.5V
4. Arcturus	34	-0.04	0.2	K1.5III
5. Vega	25	0.03	0.6	A0V
6. Capella	41	0.08	0.4	G6III + G2III
7. Rigel	860	0.12	-8.1	B8Ia
8. Procyon	11.4	0.38	2.6	F5IV
9. Achernar	69	0.46	-1.3	B3V
10. Betelgeuse	640	0.50	-7.2	M2Iab
11. Hadar	320	0.61	-4.4	B1III
12. Alpha Crucis	510	0.76	-4.6	B0.5I + B1V
13. Altair	16	0.77	2.3	A7V
14. Aldebaran	60	0.85	-0.3	K5III
15. Antares	600	0.96	-5.2	M1.5Ib
16. Spica	220	0.98	-3.2	B1V
17. Pollux	40	1.14	0.7	K0III
18. Fomalhaut	22	1.16	2.0	A3V
19. Beta Crucis	460	1.25	-4.7	B0.5III
20. Deneb	2600	1.25	-7.2	A2Ia

Constellations

Nominative	Generic ending	Abbrev.	α (h)	δ (°)	Translation
Andromeda	-dae	And	1	40 N	Chained Maiden
Antlia	-liae	Ant	10	35 S	Pump
Apus	-podis	Aps	16	75 S	Bird of Paradise
Aquarius (z)	-rii	Aqr	23	15 S	Water Bearer
Aquila	-lae	Aql	20	5 N	Eagle
Ara	-rae	Ara	17	55 S	Altar
Aries (z)	-ietis	Ari	3	20 N	Ram
Auriga	-gae	Aur	6	40 N	Charioteer
Bootes	-tis	Boo	15	30 N	Herdsman
Caelum	-aeli	Cae	5	40 S	Chisel
Camelopardalis	-di	Cam	6	70 N	Giraffe
Cancer	-cri	Cnc	9	20 N	Crab
Canes Venatici	-num -corum	CVn	13	40 N	Hunting Dogs
Canis Major	-is -ris	CMa	7	20 S	Big Dogs
Canis Minor	-is -ris	CMi	8	5 N	Little Dog
Capricornus	-ni	Cap	21	20 S	Goat
Carina	-nae	Car	9	60 S	Ship's Keel
Cassiopeia	-peiae	Cas	1	60 N	Lady in Chair
Centaurus	-ri	Cen	13	50 S	Centaur
Cepheus	-phei	Cep	22	70 N	King
Cetus	-ti	Cet	2	10 S	Whale
Chamaeleon	-ntis	Cha	11	80 S	Chameleon
Circinus	-ni	Cir	15	60 S	Compass
Columba	-bae	Col	6	35 S	Dove
Coma Berenices	-mae -cis	Com	13	20 N	Bernice's Hair
Corona Australis	-nae -lis	CrA	19	40 S	Southern Crown
Corona Borealis	-nae -lis	CrB	16	30 N	Northern Crown

Corvus	-vi	Crv	12	20 S	Crow
Crater	-eris	Crt	11	15 S	Cup
Crux	-ucis	Cru	12	60 S	Southern Cross
Cygnus	-gni	Cyg	21	40 N	Swan
Delphinus	-ni	Del	21	10 N	Dolphin
Dorado	-dus	Dor	5	65 S	Swordfish
Draco	-onis	Dra	17	65 N	Dragon
Equuleus	-lei	Equ	21	10 N	Little Horse
Eridanus	-ni	Eri	3	20 S	River Eridanus
Fornax	-acis	For	3	30 S	Furnace
Gemini	-norum	Gem	7	20 N	Twins
Grus	-ruis	Gru	22	45 S	Crane
Hercules	-lis	Her	17	30 N	Kneeling Giant
Horologium	-gii	Hor	3	60 S	Clock
Hydra	-drae	Hya	10	20 S	Water Monster
Hydrus	-dri	Hyi	2	75 S	Sea Serpent
Indus	-di	Ind	21	55 S	Indian
Lacerta	-tae	Lac	22	45 N	Lizard
Leo	-onis	Leo	11	15 N	Lion
Leo Minor	-onis -ris	LMi	10	35 N	Little Lion
Lepus	-poris	Lep	6	20 S	Hare
Libra (z)	-rae	Lib	15	15 S	Scales
Lupus	-pi	Lup	15	45 S	Wolf
Lynx	-ncis	Lyn	8	45 N	Lynx
Lyra	-rae	Lyr	19	40 N	Harp
Mensa	-sae	Men	5	80 S	Table
Microscopium	-pii	Mic	21	35 S	Microscope
Monoceros	-rotis	Mon	7	5 S	Unicorn
Musca	-cae	Mus	12	70 S	Fly
Norma	-mae	Nor	16	50 S	Level
Octans	-ntis	Oct	22	85 S	Octant
Ophiuchus	-chi	Oph	17	0	Serpent Bearer
Orion	-nis	Ori	5	5 N	Hunter
Pavo	-vonis	Pav	20	65 S	Peacock
Pegasus	-si	Peg	22	20 N	Winged Horse
Perseus	-sei	Per	3	45 N	Champion
Phoenix	-nicis	Phe	1	50 S	Phoenix
Pictor	-ris	Pic	6	55 S	Easel
Pisces (z)	-cium	Pse	1	15 N	Fish
Piscis Austrinus	-is -ni	PsA	22	30 S	Southern Fish
Puppis **	-ppis	Pup	8	40 S	Ship's Stern

Pyxis **	-xidis	Pyx	9	30 S	Ship's Compass
Reticulum	-li	Ret	4	60 S	Net
Sagitta	-tae	Sge	20	10 N	Arrow
Sagittarius (z)	-rii	Sgr	19	25 S	Archer
Scorpius (z)	-pii	Sco	17	40 S	Scorpion
Sculptor	-ris	Scl	0	30 S	Sculptor
Scutum	-ti	Sct	19	10 S	Shield
Serpens *	-ntis	Ser	18	5 S	Serpent
Sextans	-ntis	Sex	10	0	Sextant
Taurus (z)	-ri	Tau	4	15 N	Bull
Telescopium	-pii	Tel	19	50 S	Telescope
Triangulum	-li	Tri	2	30 N	Triangle
Triangulum Australe	-li -lis	TrA	16	65 S	Southern Triangle
Tucana	-nae	Tuc	0	65 S	Toucan
Ursa Major	-sae -ris	UMa	11	50 N	Big Bear
Ursa Minor	-sae -ris	UMi	15	70 N	Little Bear
Vela **	-lorum	Vel	9	50 S	Ship's Sails
Virgo (z)	-ginis	Vir	13	0	Virgin
Volans	-ntis	Vol	8	70 S	Flying Fish
Vulpecula	-lae	Vul	20	25 N	Little Fox

Astronomical/ Physical Equations

Law of Gravity

$$F_g = \frac{G\, m_1 m_2}{d^2}$$

Wien's Law

$$T\,(K) = \frac{3 \times 10^6\ nm\ K}{\lambda peak\ (in\ nm)}$$

Stefan-Boltzmann Law

$$L = 4\pi R^2 \sigma\, T^4$$

Doppler Effect
(Radial Velocity)

$$V_{rad} = \frac{\Delta\lambda\ c}{\lambda_{lab}}$$

Triangulation (distance)
(For Stars)

$$Dist.\ (in\ pc) = \frac{1}{p\ (in\ arc\ sec)}$$

Distance Modulus

$$Dist.\ (in\ pc) = 10^{(m-M+5)/5}$$

Main Sequence Lifetime

$$Life. = \frac{10^{10}\ yrs}{(X\ M\Theta)^{2.5}}$$

Gravitational Radius
(Schwarzschild Radius)

$$R_G = 2\ mi.\ per\ M\Theta$$

Astronomical Constants

Astronomical Unit (A.U.)	150,000,000 km (93,000,000 mi.)
Light Year (LY)	9.46×10^{12} km (5.88×10^{12} mi.) = 63,000 A.U.
Parsec (pc)	3.09×10^{13} km (1.92×10^{13} mi.) = 206,265 A.U.
Sidereal Year	365.2564 days
Tropical Year	365.2422 days
Gregorian Year	365.2425 days
Earth's Mass	5.9736×10^{24} kg
Sun's Mass	1.9891×10^{30} kg = 333,000 Earth masses
Earth's Diameter	6371 km (7918 mi.)
Sun's Diameter	1.39×10^{6} km (8.65×10^{5} mi.) = 109 Earth diameters
Sun's Luminosity	3.827×10^{26} watts

Physical Constants

Speed of Light (c)	300,000 km/sec (186,000 mi./sec)
Gravitational Constant (G)	6.67×10^{-11} m^3 /(kg sec^2)
Boltzmann Constant (k)	1.38×10^{-23} Joules/Kelvin
Stefan-Boltzmann Constant (σ)	5.67051×10^{-8} J/(m^2 K^4 s)
Wien's Law Constant	2.90×10^{6} nm K
Planck Constant (h)	6.63×10^{-34} Joules second
Electron's Mass	9.11×10^{-28} g
Proton's Mass	1.6727×10^{-24} g
Neutron's Mass	1.6750×10^{-24} g
Deuterium Nucleus' Mass	3.34×10^{-24} g